V I S U A L
D I C T I O N A R Y
ANIMALS

VISUAL
DICTIONARY
ANIMALS

FOG CITY PRESS

Published by Fog City Press
814 Montgomery Street
San Francisco, CA 94133 USA

First printed 2004

ISBN 1 877019 71 2

Color reproduction by Colourscan Co Pte Ltd
Printed by Kyodo Printing Co Pte Ltd
Printed in Singapore

A Weldon Owen Production

FOG CITY PRESS
Chief Executive Officer: John Owen
President: Terry Newell
Publisher: Lynn Humphries
Creative Director: Sue Burk
Project Editor: Jessica Cox
Project Designer: Heather Menzies
Editorial Coordinator: Jennifer Losco
Consultants: George McKay, Richard Vogt,
    Chris MacDonald, Hugh Dingle
Production Manager: Caroline Webber
Production Coordinator: James Blackman
Sales Manager: Emily Jahn
Vice President International Sales: Stuart Laurence

Produced using arkiva retrieval technology
For further information, contact arkiva@weldonowen.com.au

Contents

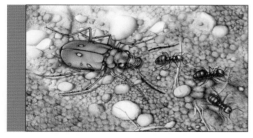

Living World

WORLD CLIMATES

- Tropical
- Subtropical
- Desert and semi-desert
- Dry temperate
- Wet temperate
- Cold temperate
- Polar
- Mountain
- → Cool currents
- → Warm currents

Oceans
The oceans' climates change with latitude, and warm and cool currents.

Tropical rainforests
Tropical climates and rainforests are found in equatorial regions.

Subtropical savannas
Savanna grasslands are found across tropical and subtropical latitudes.

Deserts and semi-deserts
Many arid zones lie downwind of mountains, resulting in dry conditions.

DIVERSE ECOSYSTEMS
Planet Earth's weather systems have created
a range of habitats. Within each habitat is
a community of plants and animals, called
an ecosystem. To survive, individuals must
adapt to their environment, making the
most of the resources it has to offer.

Alpine and polar regions
These regions are typified
by their harsh climates
and barren landscapes.

Temperate grasslands
Dry temperate climates
are found in mid-latitudes,
and produce grasslands.

Temperate forests
Wet temperate climates
are found in mid-latitudes,
and produce forests.

Coniferous forests
Northern temperate (or
boreal) climates occur in
the northern hemisphere.

Changing Habitats

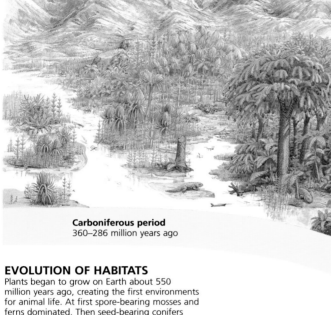

Carboniferous period
360–286 million years ago

Permian period
286–248 million years ago

Triassic period
248–208 million years ago

EVOLUTION OF HABITATS
Plants began to grow on Earth about 550 million years ago, creating the first environments for animal life. At first spore-bearing mosses and ferns dominated. Then seed-bearing conifers and flowering plants emerged. As plants evolved so did the animals that ate them, resulting in the complex ecosystems of today.

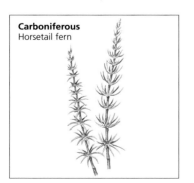

Carboniferous
Horsetail fern

Permian
Ginko

Triassic
Cyclad

Jurassic
Conifer

Cretaceous
Magnolia

Tertiary to modern day
Strawberry

Jurassic period
208–144 million years ago

Cretaceous period
144–65 million years ago

Tertiary period
65 million years ago

Tropical Rainforests

BROMELIAD LIFE

Velvet worms can be found here, high above their usual forest-floor home.

Mosquitoes breed in the bromeliad pool.

Female poison frog deposits tadpoles into water-filled bromeliads.

Alligator lizard drinks from the pool, safe above the forest floor.

Beetle larvae prey on poison frog tadpoles.

Land snails favor the moist leaves for foraging and hiding.

Mouse opossum eats a damselfly larva from the bromeliad pool.

RAINFOREST PLANTS

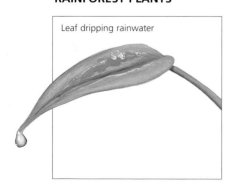

Leaf dripping rainwater

Flowering ginger

Epiphytic plant

Savannas

SAVANNA PLANTS

Sturt's pea

Wheat

Eucalyptus flower

GELADA BABOON HABITAT

Colored chest markings alter with a female's state of fertility.

Gelada baboons live in large social groups on the high grasslands of Ethiopia and Eritrea.

They are the only species of primate to eat grasses, which are supplemented by roots, stems and fruits.

SUBTROPICAL CLIMATE
Subtropical areas generally receive rainfall only in summer. This provides perfect conditions for savanna grasslands rather than dense forests.

Deserts

DESERT BY DAY

Plant and animal species that live in deserts have adapted to the lack of water. Some plants store water in their stems; others have long roots reaching underground.

DESERT LIFE

Jeweled gecko feeds on invertebrates in the spinifex, emitting a toxic fluid when startled.

Crimson chats live in open plains with a cover of spinifex or saltbush.

Small nocturnal cockroaches feed on decaying organic matter.

Ningauis are tiny marsupials that quench their thirst by licking dew from spinifex.

Burton's legless lizard preys on smaller lizards, suffocating and then swallowing them.

Desert skink eats termites and other invertebrates in the spinifex community.

DESERT BY NIGHT

Because of the intense heat of the day in some regions, many desert animals are nocturnal. Some plants have adapted to flower at night so they will be pollinated.

Beaver-tail cactus

Quiver tree

Century plant flowers

Chain-fruit cholla

21

Temperate Regions

TEMPERATE CLIMATE

Mid-latitude areas are typified by four distinct seasons. Depending on rainfall, temperate regions can be grasslands, shrublands or woodlands. To survive cold winters, animals may migrate or hibernate.

Spring: leaves sprout; flowers bloom.

FOUR SEASONS

Summer: fertilized flowers form fruits.

Autumn: fruits seed; leaves fall.

Winter: buds protect new growth.

TEMPERATE LIFE

Pika

Cuckoo

Cattle

Woodlands

WOODLAND PLANTS

Autumn leaves

Pine cones

Maple seeds

CONIFEROUS FORESTS

Evergreen conifers thrive in the forests of cold temperate regions. The trees are cone-shaped, allowing snow to slide off. Many animals have thick fur for warmth, while some hibernate to escape the bitter cold.

Gamebirds such as grouse have thick, downy plumage to survive the cold winters.

With coloring like mottled bark, owls are well suited to coniferous forests.

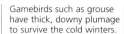

Wolverines are skilled hunters in deep snow. Their fur is hollow for insulation.

Solitary mooses, the largest of all deer, gather in groups only in the breeding season.

To survive winter, squirrels gather a cache of nuts and berries every autumn.

Alpine Regions

ALPINE PLANTS

Edelweiss

Puya

Conifers

HYRAX HABITAT

Rock and bush hyraxes live in close proximity.

Rock hyraxes feed on alpine grasses.

Bush hyraxes forage for leaves and bark.

ALPINE CLIMATE

As altitudes and latitudes increase, vegetation decreases, until trees can no longer grow. Mountain areas are similar to arctic tundra: plants and animals adapt to flourish in short summers and survive long winters.

Seashores

COASTAL PLANTS

Coconut sprouting

Screw pine fruit

Casuarina leaves

SEASHORES

Only salt-resistant plants can grow close to the sea, with grasses and shrubs forming the frontline against sea breezes. Below the splash line, tidal plants and animals thrive in the pools and among rocks.

Hermit crabs protect their soft bodies in discarded mollusk shells.

Mussels grow in dense colonies attached to rocks.

Clinging to rocks like mollusks, barnacles are actually crustaceans.

Blennies hide under stones, darting from crevice to shadow.

Carnivorous starfish move slowly on hundreds of feet.

Small octopuses are secretive inhabitants of tidepools.

Seaweeds are marine algae, and lack a plant's roots, flowers, seeds and fruit.

Colorful sea anemones feed on small animals stung by their tentacles.

Sculpin are predators that lie in wait for prey rather than chase it.

Crabs are abundant in intertidal tidepools.

Oceans

OCEAN HABITATS

The oceans contain many habitats, dependent on latitude and depth. A huge variety of animals and plants have adapted to life underwater, from seaweeds and seagrasses, to sharks, corals and fishes.

Dugongs live among the coastal seagrass beds in the Indo-Pacific region.

Yellow-finned leatherjackets are herbivorous fishes.

Blue swimmer-crabs bury themselves in sand up to their eyes to wait for prey.

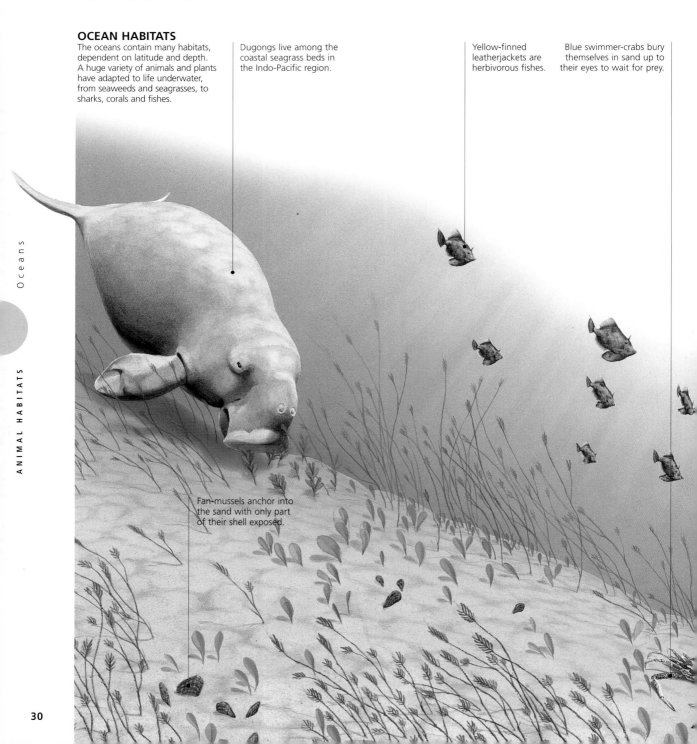

Fan-mussels anchor into the sand with only part of their shell exposed.

OCEAN LIFE

Solitary green turtles migrate vast distances to lay their eggs.

Sunlight zone
Enough sunlight penetrates this layer for plants to live. It contains most of the ocean's life.

Twilight zone
Not enough sunlight to sustain plant life. Many fishes in this zone contain light-producing bacteria and glow in the dark.

Midnight zone
This is a vast mass of cold, slow-moving water in absolute darkness. Life is about a tenth that of the twilight zone and is also less diverse.

Abyssal zone
The near-freezing, pitch-black bottom layer has few species and little food. Near hydrothermal vents, some animals make food using internal bacteria and hydrogen sulfide.

0

650 ft
(200 m)

3,250 ft
(1,000 m)

9,850 ft
(3,000 m)

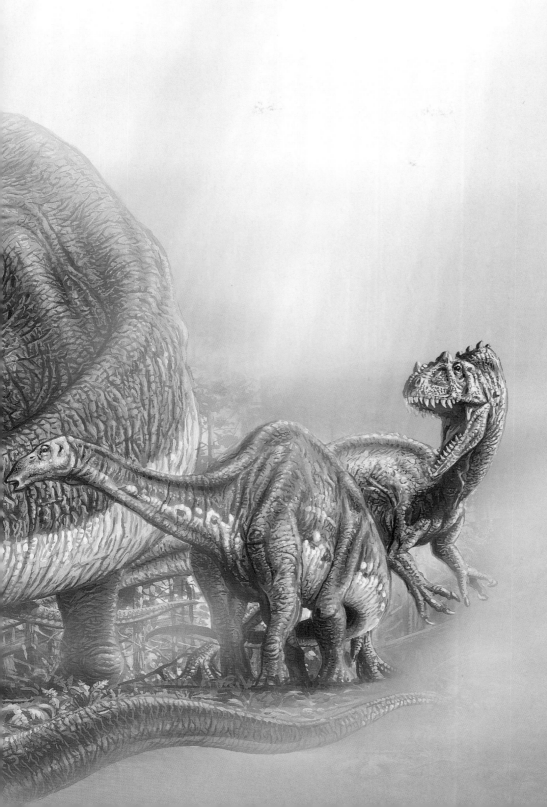

Web of Life

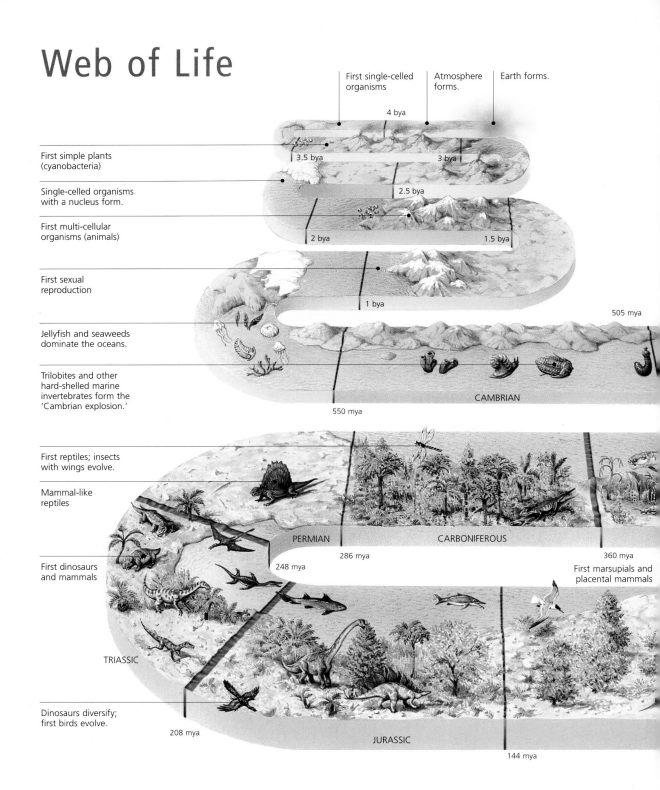

First single-celled organisms

Atmosphere forms.

Earth forms.

4 bya

First simple plants (cyanobacteria)

3.5 bya

3 bya

Single-celled organisms with a nucleus form.

2.5 bya

First multi-cellular organisms (animals)

2 bya

1.5 bya

First sexual reproduction

1 bya

505 mya

Jellyfish and seaweeds dominate the oceans.

Trilobites and other hard-shelled marine invertebrates form the 'Cambrian explosion.'

CAMBRIAN

550 mya

First reptiles; insects with wings evolve.

Mammal-like reptiles

PERMIAN

CARBONIFEROUS

286 mya

360 mya

248 mya

First dinosaurs and mammals

First marsupials and placental mammals

TRIASSIC

Dinosaurs diversify; first birds evolve.

208 mya

JURASSIC

144 mya

FOSSIL RECORD

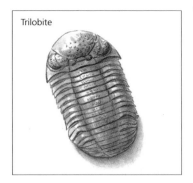

Trilobite

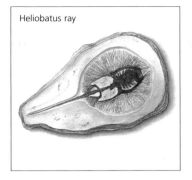

Heliobatus ray

Snakefly

Dinosaur embryo

Archean era	Cenozoic era
Proterozoic era	— Mass extinction
Paleozoic era	**bya** billion years ago
Mesozoic era	**mya** million years ago

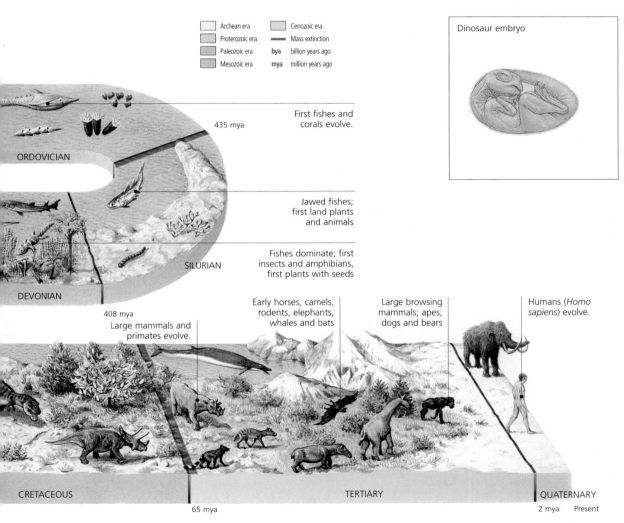

ORDOVICIAN

435 mya

First fishes and corals evolve.

SILURIAN

Jawed fishes; first land plants and animals

DEVONIAN

Fishes dominate; first insects and amphibians, first plants with seeds

408 mya
Large mammals and primates evolve.

Early horses, camels, rodents, elephants, whales and bats

Large browsing mammals; apes, dogs and bears

Humans (*Homo sapiens*) evolve.

CRETACEOUS

65 mya

TERTIARY

QUATERNARY

2 mya   Present

Before the Dinosaurs

Ornithosuchus | Dimetrodon | Dragonfly | Hylonomus | Ichthyostega | Dunkleosteus

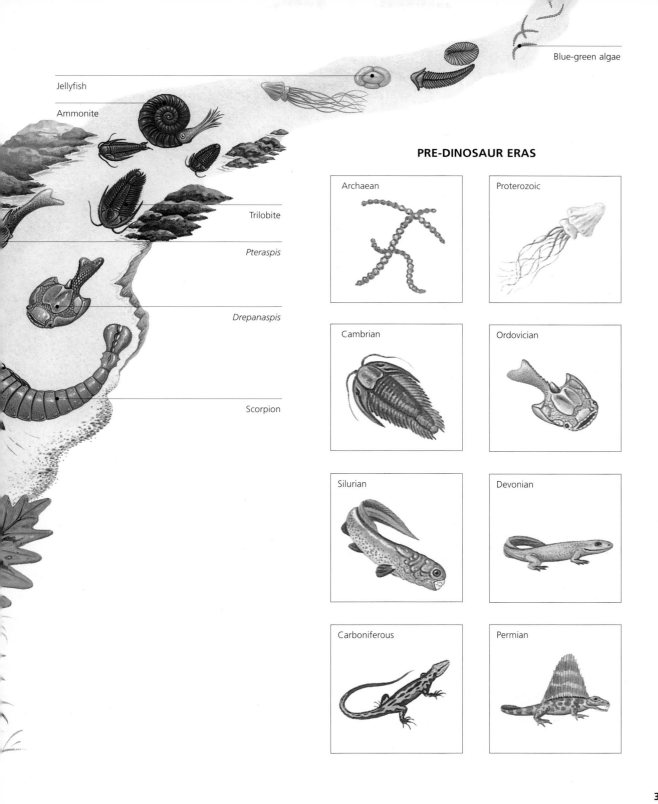

Blue-green algae

Jellyfish

Ammonite

Trilobite

Pteraspis

Drepanaspis

Scorpion

PRE-DINOSAUR ERAS

Archaean

Proterozoic

Cambrian

Ordovician

Silurian

Devonian

Carboniferous

Permian

Triassic Period

Triassic landmasses

TRIASSIC LANDSCAPE

NOTHOSAURUS

EUDIMORPHODON

HERRERASAURUS

PROCOMPSOGNATHUS

SELLOSAURUS

SALTOPUS

MELANOROSAURUS

KANNEMEYERIA

TRIASSIC DINOSAUR DIETS

Dragonfly

Wielandiella

Haramiya

Tree fern

Jurassic Period

Jurassic landmasses

JURASSIC LANDSCAPE

STEGOSAURUS

ALLOSAURUS

CAMPTOSAURUS

COELURUS

CERATOSAURUS

RHAMPHORHYNCHUS

HETERODONTOSAURUS

DIPLODOCUS

JURASSIC DINOSAUR DIETS

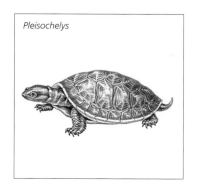

Pleisochelys

Archaeopteryx

Cockroach

Cretaceous Period

Cretaceous landmasses

CRETACEOUS LANDSCAPE

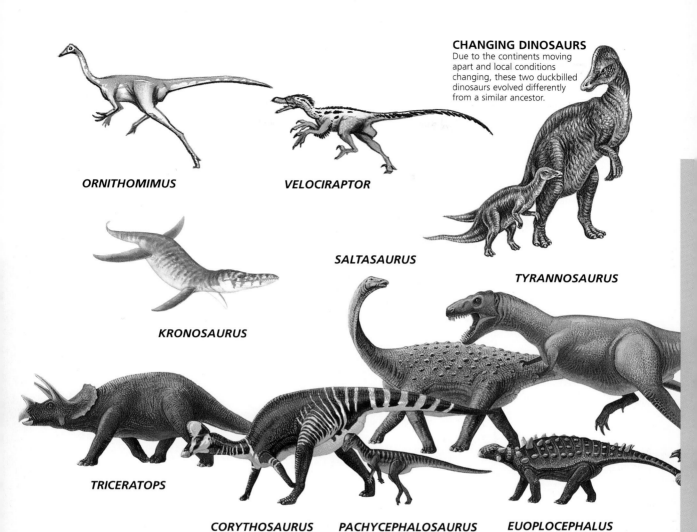

ORNITHOMIMUS

VELOCIRAPTOR

CHANGING DINOSAURS
Due to the continents moving apart and local conditions changing, these two duckbilled dinosaurs evolved differently from a similar ancestor.

SALTASAURUS

TYRANNOSAURUS

KRONOSAURUS

TRICERATOPS

CORYTHOSAURUS

PACHYCEPHALOSAURUS

EUOPLOCEPHALUS

CRETACEOUS LIFE

Polyglyphanodon

Magnolia

Crusafontia

Types of Dinosaurs

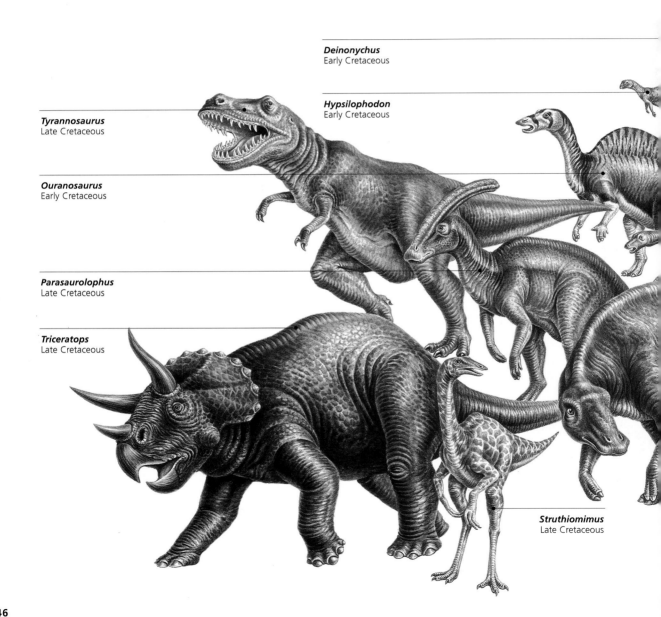

Deinonychus
Early Cretaceous

Hypsilophodon
Early Cretaceous

Tyrannosaurus
Late Cretaceous

Ouranosaurus
Early Cretaceous

Parasaurolophus
Late Cretaceous

Triceratops
Late Cretaceous

Struthiomimus
Late Cretaceous

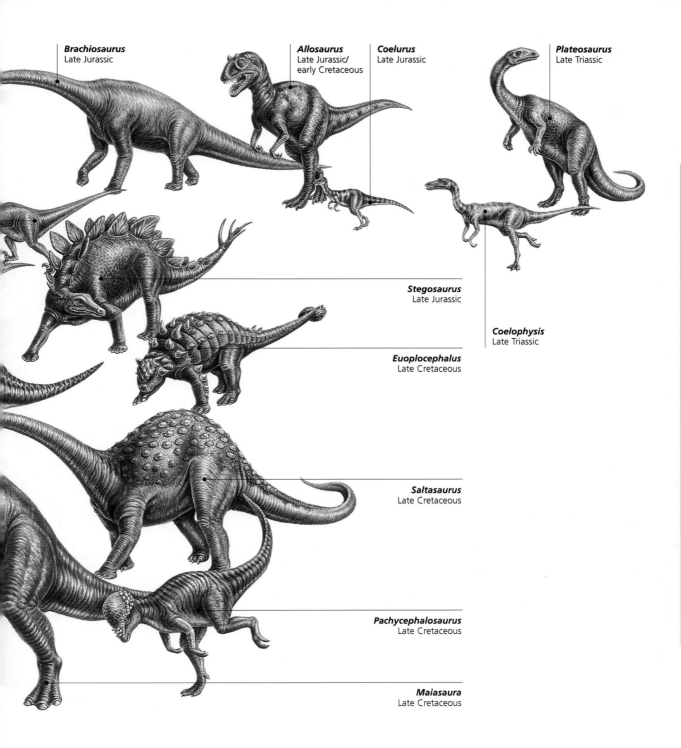

Brachiosaurus
Late Jurassic

Allosaurus
Late Jurassic/
early Cretaceous

Coelurus
Late Jurassic

Plateosaurus
Late Triassic

Stegosaurus
Late Jurassic

Coelophysis
Late Triassic

Euoplocephalus
Late Cretaceous

Saltasaurus
Late Cretaceous

Pachycephalosaurus
Late Cretaceous

Maiasaura
Late Cretaceous

Dinosaur Characteristics

DINOSAUR ANTATOMY

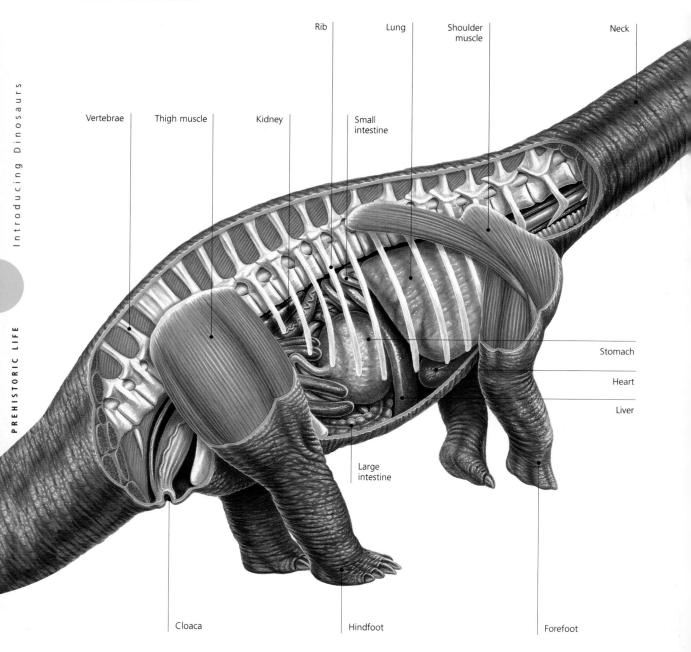

Rib

Lung

Shoulder muscle

Neck

Vertebrae

Thigh muscle

Kidney

Small intestine

Stomach

Heart

Liver

Large intestine

Cloaca

Hindfoot

Forefoot

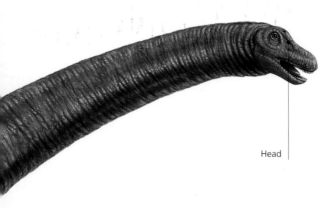

Head

BRAIN COMPARISON

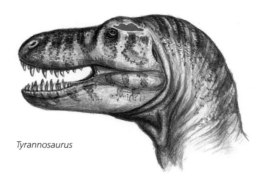

Tyrannosaurus

Troödon

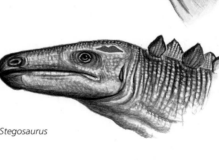

Stegosaurus

Human

TYPES OF TEETH

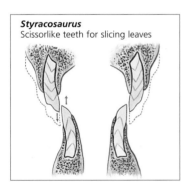

Styracosaurus
Scissorlike teeth for slicing leaves

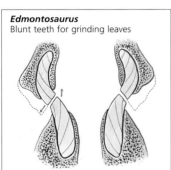

Edmontosaurus
Blunt teeth for grinding leaves

49

Dinosaur Hips

EVOLUTION OF HIPS

Legs sprawling on either side of body

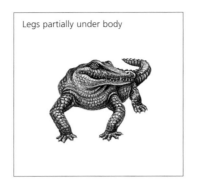

Legs partially under body

Legs fully under body

Pubis points backward; allows space for large intestines of quadrapedal (four-legged) plant eaters.

EDMONTOSAURUS
Bird-hipped dinosaur

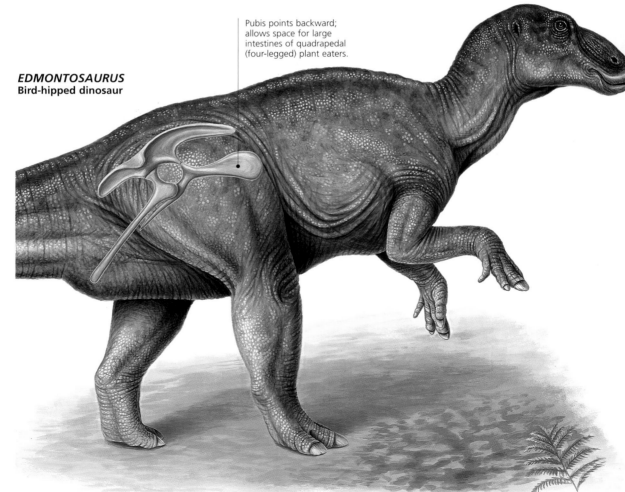

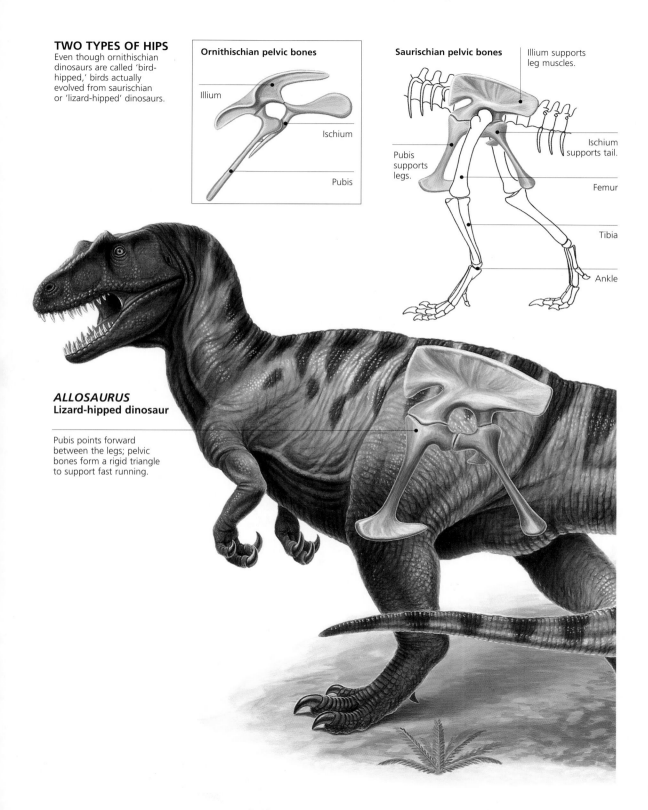

TWO TYPES OF HIPS

Even though ornithischian dinosaurs are called 'bird-hipped,' birds actually evolved from saurischian or 'lizard-hipped' dinosaurs.

Ornithischian pelvic bones

Illium

Ischium

Pubis

Saurischian pelvic bones

Illium supports leg muscles.

Pubis supports legs.

Ischium supports tail.

Femur

Tibia

Ankle

ALLOSAURUS
Lizard-hipped dinosaur

Pubis points forward between the legs; pelvic bones form a rigid triangle to support fast running.

Raising Young

HATCHING OUT

Maiasaura hatchling

One-year-old Maiasaura

Adult Maiasaura

INSIDE AN EGG

Allantois stored waste.

Membrane (chorion) provided oxygen.

Albumen (yolk sac) nourished embryo.

Eggshell protected embryo.

Amniotic (fluid) sac cushioned dinosaur.

OVIRAPTOR NEST

Mother *Oviraptor* feeds a freshly killed young *Velociraptor* to her nest of hatchlings. Unlike most modern birds and reptiles, some dinosaurs cared for their young after hatching.

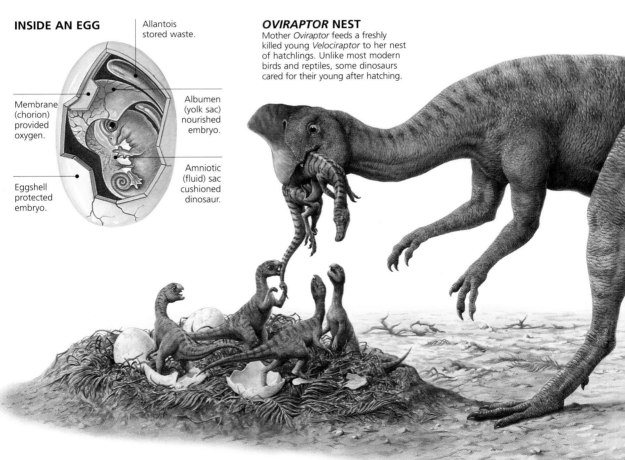

DINOSAUR NEST

Up to 25 eggs were laid in each *Maiasaura* nest.

Nests were deep bowls dug from the surrounding mud.

Nests were 7 feet (2 m) wide and 3 feet (1 m) deep.

Hatchlings were cared for until ready to leave the nest.

Maiasaura hatchlings were about 1½ feet (50 cm) long at birth.

EGG ARRANGEMENTS

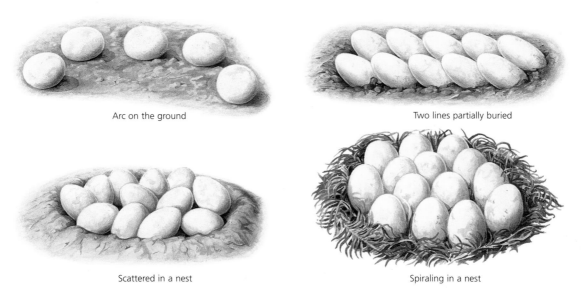

Arc on the ground

Two lines partially buried

Scattered in a nest

Spiraling in a nest

Sauropods

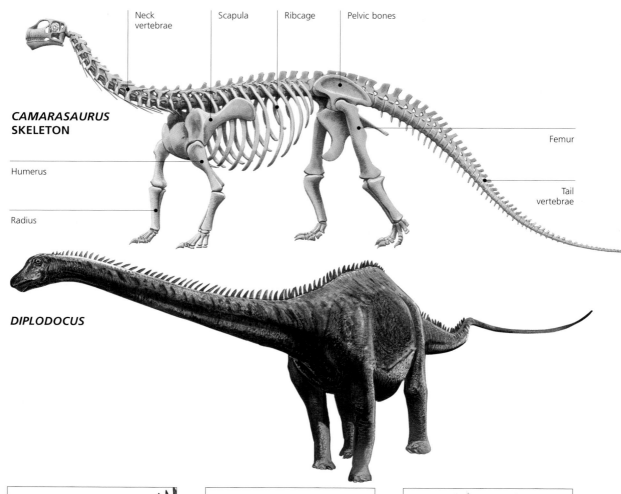

Neck
vertebrae

Scapula

Ribcage

Pelvic bones

**CAMARASAURUS
SKELETON**

Humerus

Radius

Femur

Tail
vertebrae

DIPLODOCUS

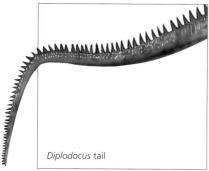

Diplodocus tail

Diplodocus tail vertebra

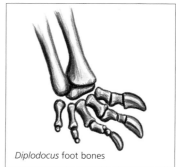

Diplodocus foot bones

PLATEOSAURUS

BRACHIOSAURUS

Plateosaurus skull

Brachiosaurus head

Brachiosaurus skull

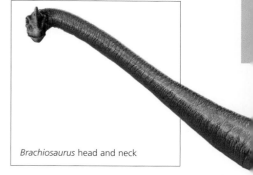

Brachiosaurus head and neck

Theropods

STRUTHIOMIMUS SKELETON

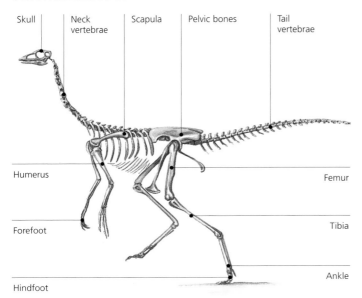

Skull | Neck vertebrae | Scapula | Pelvic bones | Tail vertebrae

Humerus

Forefoot

Hindfoot

Femur

Tibia

Ankle

EATING TOOLS

Allosaurus skull
Hinged jaws to swallow prey whole

Carcharodontosaurus teeth
Serrated teeth front and back to slice flesh

STRUTHIOMIMUS

ADAPTATIONS FOR SURVIVAL

Tyrannosaurus had hinged jaws and backward-facing teeth to rip and tear flesh.

Gallimimus ran at speeds of up to 30 miles (48 km) an hour. A slower dinosaur such as *Albertosaurus* had no chance of catching it.

FEEDING

Baryonyx

Oviraptor

Albertosaurus

Compsognathus

Theropods

VELOCIRAPTOR

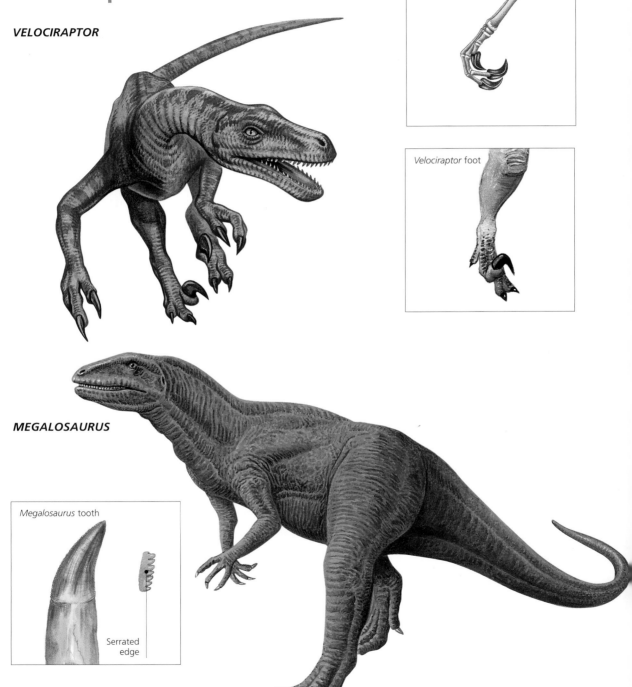

Velociraptor arm bones

Velociraptor foot

MEGALOSAURUS

Megalosaurus tooth

Serrated edge

ALLOSAURUS

Allosaurus skull

OVIRAPTOR

Two Oviraptor heads showing crests

Ornithopods

OURANOSAURUS SKELETON

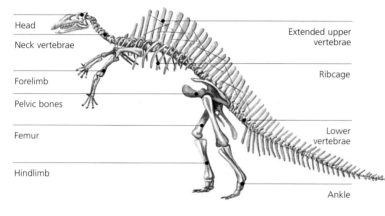

Head

Neck vertebrae

Forelimb

Pelvic bones

Femur

Hindlimb

Extended upper
vertebrae

Ribcage

Lower
vertebrae

Ankle

ORNITHOPOD SKULLS

Ouranosaurus skull

Anatotitan skull

OURANOSAURUS

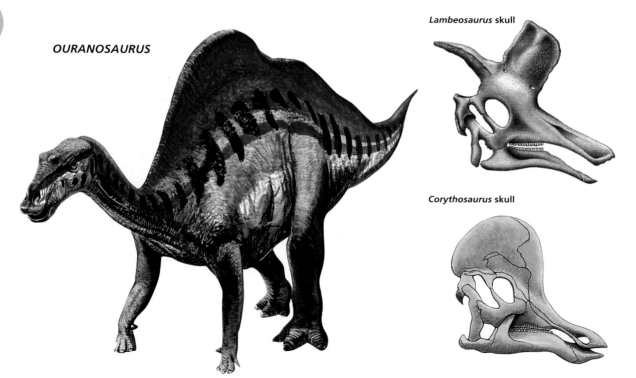

Lambeosaurus skull

Corythosaurus skull

IGUANODONS

PARASAUROLOPHUS

Juvenile

Adult female

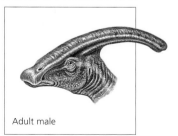

Adult male

Armored Dinosaurs

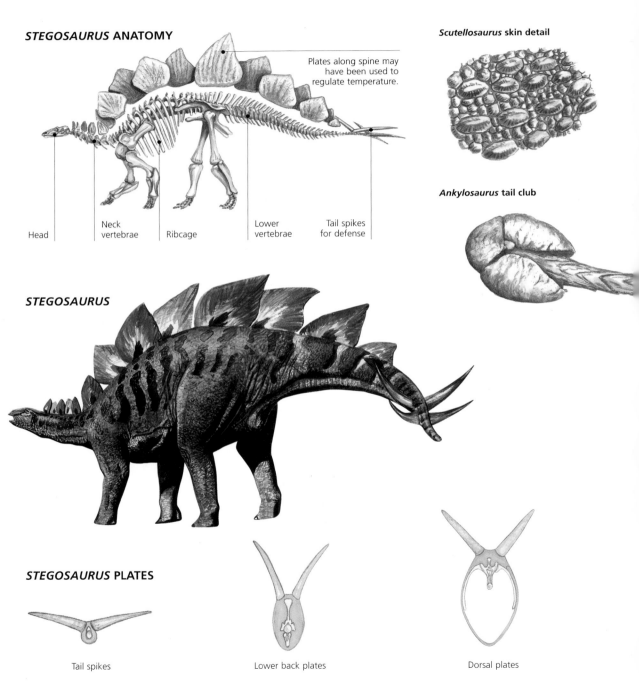

STEGOSAURUS ANATOMY

Plates along spine may have been used to regulate temperature.

Head

Neck vertebrae

Ribcage

Lower vertebrae

Tail spikes for defense

STEGOSAURUS

STEGOSAURUS PLATES

Tail spikes

Lower back plates

Dorsal plates

Scutellosaurus skin detail

Ankylosaurus tail club

TUOJIANGOSAURUS

Tuojiangosaurus tail
Spikes used to stab
potential predators

EUOPLACEPHALUS

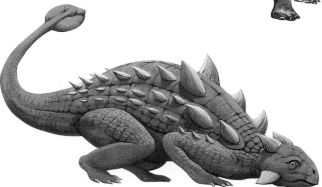

Euoplacephalus tail
Clublike bony tail
used to hit predators

Euoplacephalus skull
Armor-plated with
spikes to protect neck

EDMONTONIA

Edmontonia head
Forward-facing spikes
used to ram predators

Ceratopians

CHASMOSAURUS HEAD

CHASMOSAURUS

STYRACOSAURUS SKULL

STYRACOSAURUS

TRICERATOPS

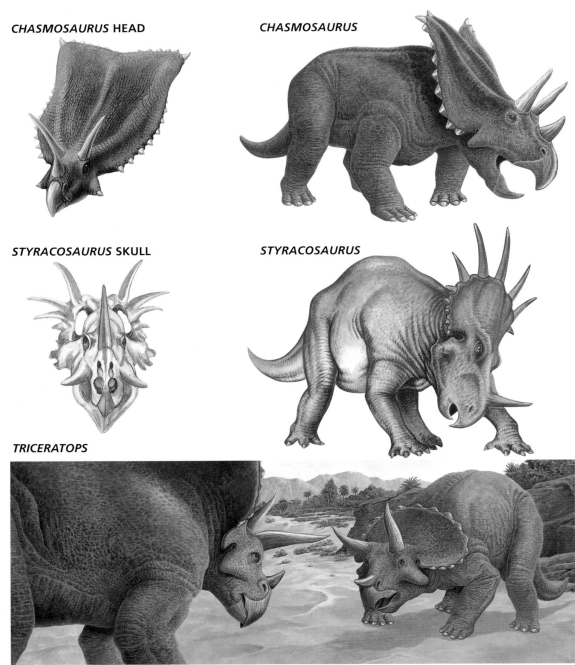

PROTOCERATOPS

PROTOCERATOPS SKULL

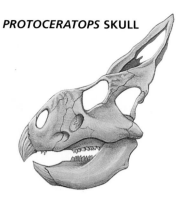

TOROSAURUS SKULL

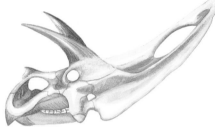

TOROSAURUS

CENTROSAURUS

CENTROSAURUS HEAD

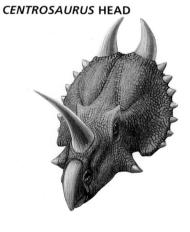

Dinosaur Contemporaries

DIMORPHODON

PTERODAUSTRO

ARCHAEOPTERYX

DIMETRODON

MESONYX

ALPHADON

DORUDON

REPTILES OF THE SEA

ICHTHYOSAURUS

LIOPLEURODON

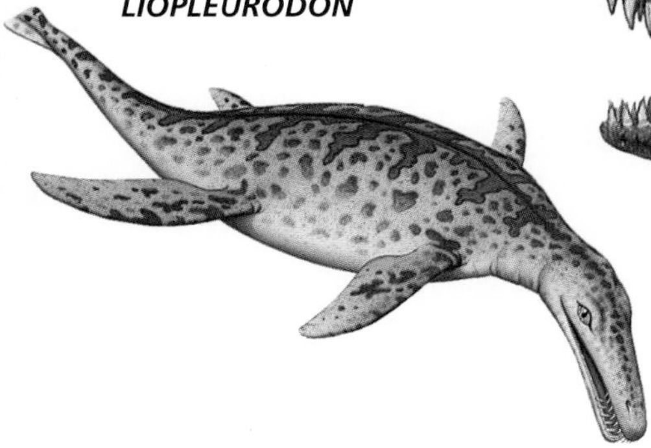

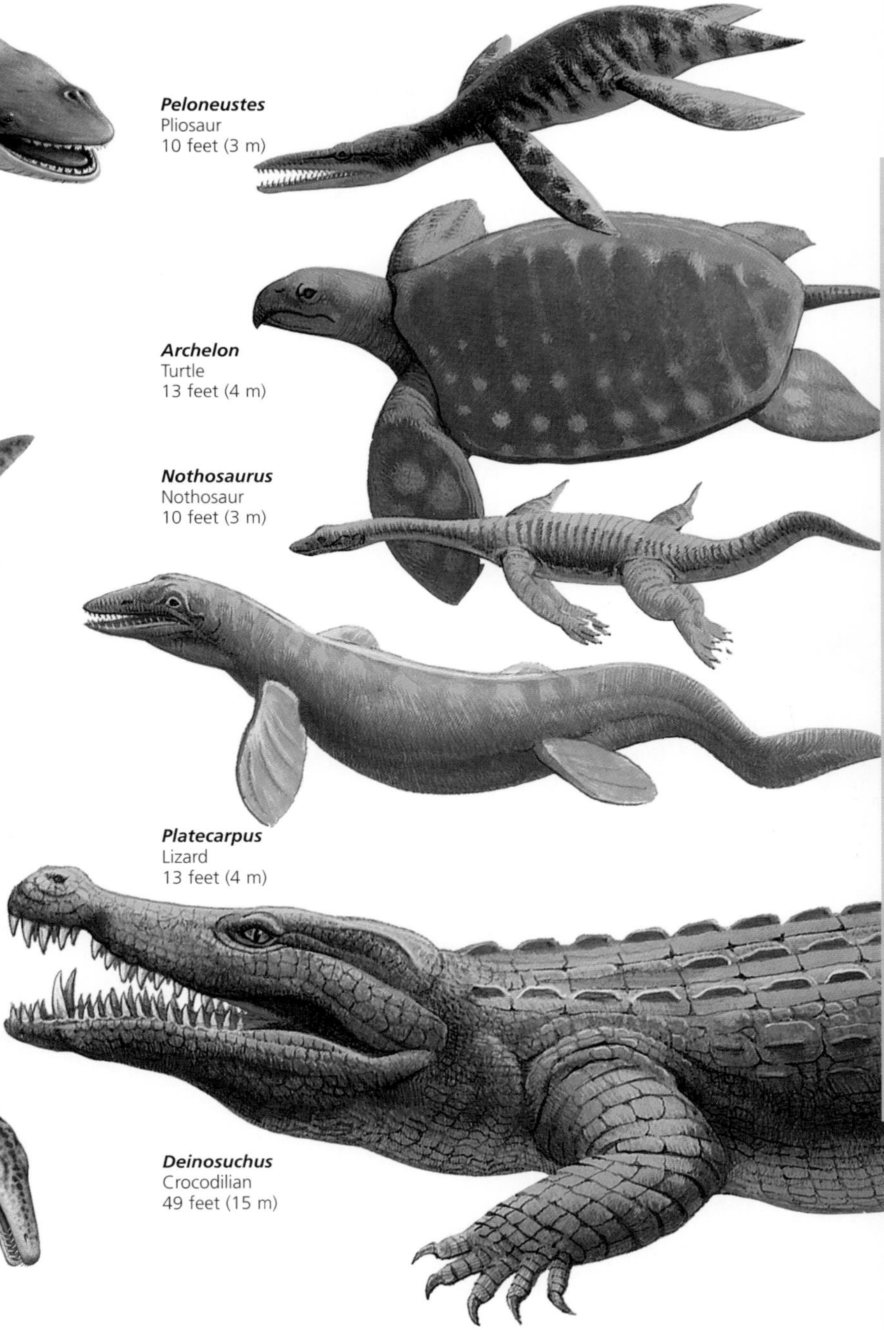

Peloneustes
Pliosaur
10 feet (3 m)

Archelon
Turtle
13 feet (4 m)

Nothosaurus
Nothosaur
10 feet (3 m)

Platecarpus
Lizard
13 feet (4 m)

Deinosuchus
Crocodilian
49 feet (15 m)

Dinosaur Extinction

EXTINCTION THEORIES

Hot climate change

Cold climate change

Volcanic eruptions

Meteorite crashing to Earth

DINOSAUR EXTINCTION

There are several theories about what caused the dinosaurs to suddenly disappear 65 million years ago. The most widely accepted theory is that a huge meteorite hit Earth causing environmental chaos.

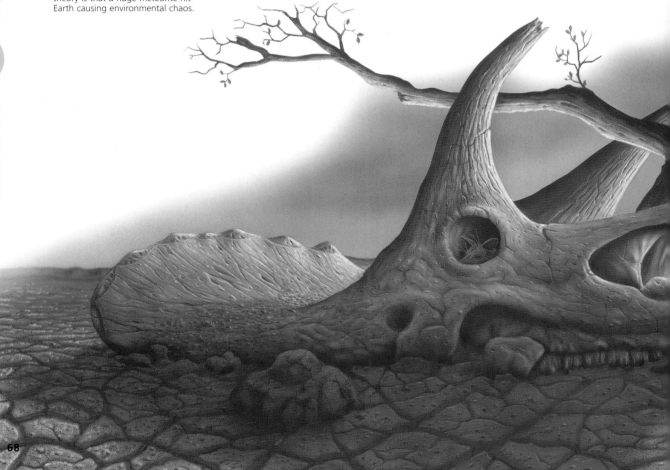

Small mammal

AGE OF MAMMALS
The small mammals of the Cretaceous were some of the species that survived the mass extinctions of 65 million years ago. These evolved and diversified into the thousands of known species today.

Recreating the Dinosaurs

FOSSIL FORMATION

Safe from scavengers
A dead dinosaur lies decomposing at the bottom of a lake, its skeleton remaining intact.

Turning to stone
Trapped and flattened by layers of sedimentary rock, dinosaur bones are replaced by minerals.

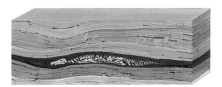

Covered up
Layers of sand or silt cover the dinosaur's bones, which stops them from being washed away.

Fossil uncovered
Millions of years later, movements within the Earth bring the fossilized skeleton to the surface.

Pieces of fossilized tail vertebrae

Missing sections of vertebrae are sketched from surrounding bones.

Sketches are painstakingly executed.

WALKING ON THE PAST

DINOSAUR BONES

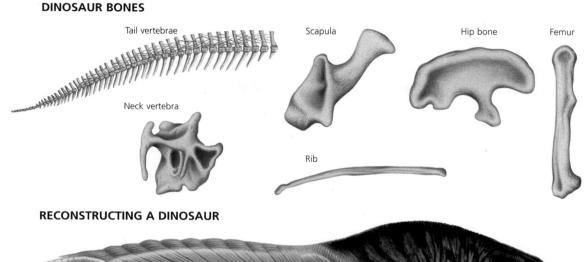

Tail vertebrae

Neck vertebra

Scapula

Hip bone

Femur

Rib

RECONSTRUCTING A DINOSAUR

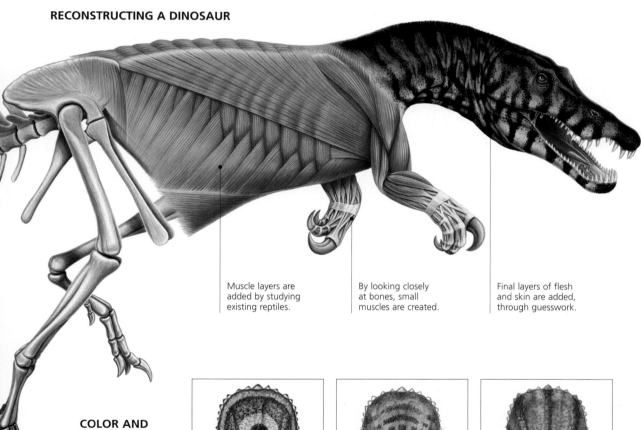

Muscle layers are added by studying existing reptiles.

By looking closely at bones, small muscles are created.

Final layers of flesh and skin are added, through guesswork.

COLOR AND TEXTURE
Because no one knows what dinosaurs actually looked like, illustrators have to guess what the color and texture of dinosaur skin was like.

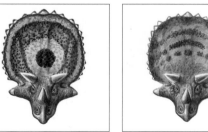

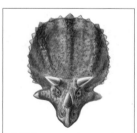

Modern Relatives

ARCHAEOPTERYX

It is generally believed that modern birds evolved from dinosaurs. The skeleton of *Archaeopteryx*, the earliest known bird, is very similar to certain dinosaurs—such that many now classify *Archaeopteryx* as a dinosaur with feathers.

FROM DINOSAUR TO BIRD

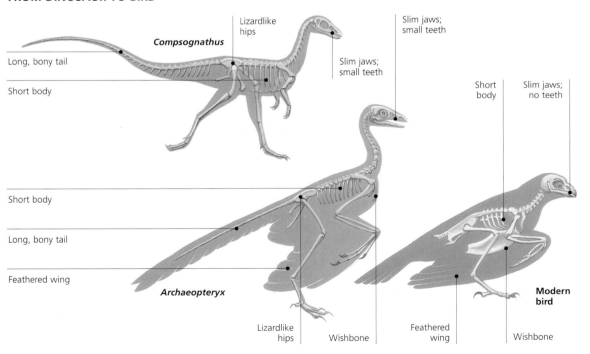

Long, bony tail

Short body

Compsognathus

Lizardlike hips

Slim jaws; small teeth

Slim jaws; small teeth

Short body

Slim jaws; no teeth

Short body

Long, bony tail

Feathered wing

Archaeopteryx

Lizardlike hips

Wishbone

Feathered wing

Wishbone

Modern bird

Waterfowl

Turtle

Crocodile

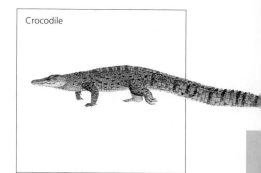

Fish

Emu

Seabird

Hoatzin

Lizard

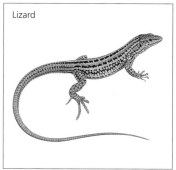

Insect

Archaeopteryx wing

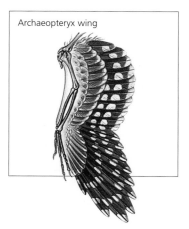

Bat wing

Pigeon wing

Classifying Mammals

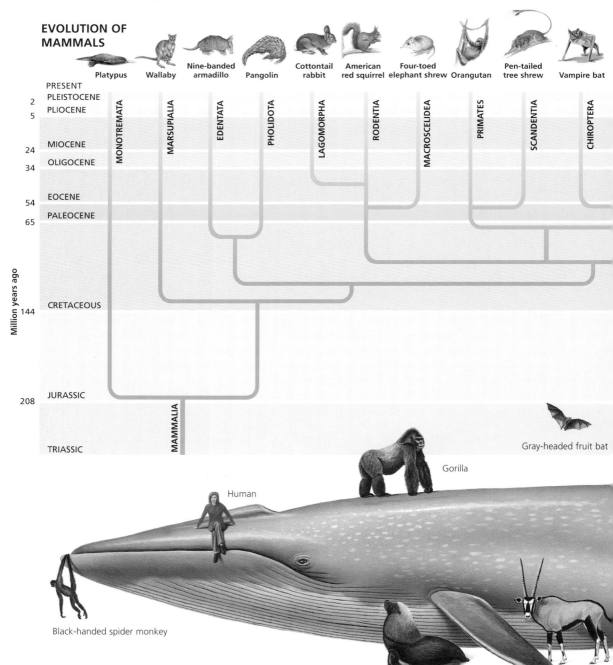

EVOLUTION OF MAMMALS

Introducing Mammals

MAMMALS

Platypus · Wallaby · Nine-banded armadillo · Pangolin · Cottontail rabbit · American red squirrel · Four-toed elephant shrew · Orangutan · Pen-tailed tree shrew · Vampire bat

Million years ago

PRESENT
PLEISTOCENE
2
5 PLIOCENE
MIOCENE
24
34 OLIGOCENE
EOCENE
54
PALEOCENE
65
CRETACEOUS
144
JURASSIC
208
TRIASSIC

MONOTREMATA · MARSUPIALIA · EDENTATA · PHOLIDOTA · LAGOMORPHA · RODENTIA · MACROSCELIDEA · PRIMATES · SCANDENTIA · CHIROPTERA

MAMMALIA

Gray-headed fruit bat

Gorilla

Human

Black-handed spider monkey

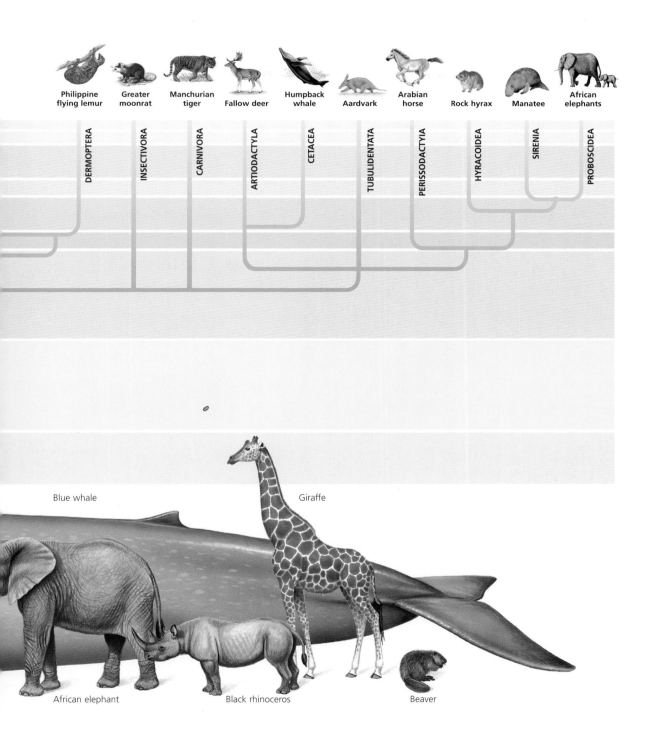

Philippine flying lemur

DERMOPTERA

Greater moonrat

INSECTIVORA

Manchurian tiger

CARNIVORA

Fallow deer

ARTIODACTYLA

Humpback whale

CETACEA

Aardvark

TUBULIDENTATA

Arabian horse

PERISSODACTYIA

Rock hyrax

HYRACOIDEA

Manatee

SIRENIA

African elephants

PROBOSCIDEA

Blue whale

Giraffe

African elephant

Black rhinoceros

Beaver

Mammal Characteristics

HUNTERS AND PREY

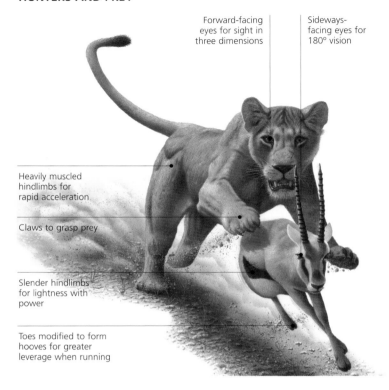

Forward-facing eyes for sight in three dimensions

Sideways-facing eyes for 180° vision

Heavily muscled hindlimbs for rapid acceleration

Claws to grasp prey

Slender hindlimbs for lightness with power

Toes modified to form hooves for greater leverage when running

MAMMAL FOREFEET

Whale: modified flipper for swimming

Bear: adapted for digging and hunting

Bat: fingers stretched for flying

GIVING BIRTH TO LIVE YOUNG

All mammals are vivaporous: they give birth to live young.

Wildebeest birth
Social mammals time mating so their young are born together, to protect them from predators.

After birth
Newly born wildebeests are able to stand and run within minutes of birth.

MAMMAL EAR

The three bones that make up a mammal's middle ear work together to transmit sound waves into the sensory cells of the inner ear.

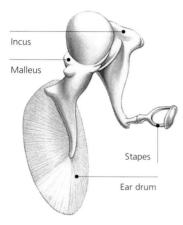

Incus

Malleus

Stapes

Ear drum

MAMMAL HAIR

All mammals have hair on their bodies. A typical hair contains three layers: an insulating medulla, cortex containing pigment and outer cuticle.

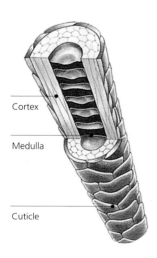

Cortex

Medulla

Cuticle

CONVERGENT EVOLUTION

Animal species that are not closely related often adapt to environmental conditions or pressures in similar ways. This is called convergent evolution.

Long nose, sticky tongue and no teeth to eat insects

Short-beaked echidna

Pangolin

Long, narrow finger to hook grubs

Aye-aye

Striped possum

Slow moving to conserve energy

Koala

Sloth

Evolution of Mammals

DIMETRODON
Mammal-like reptile that
lived 300 million years ago.

MEGAZOSTRODON
Earliest known true mammal,
it lived 220 million years ago.

CYNOGNATHUS
Mammal-like reptiles that lived
245–230 million years ago.

MORGANUCODONTID
One of the earliest true mammals,
it lived 195 million years ago.

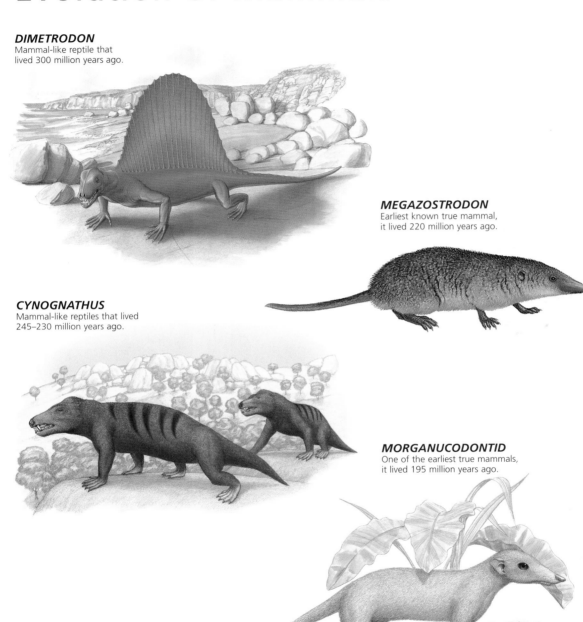

UINTATHERIUM
One of the earliest herbivores, it lived 60 million years ago.

ARSINOITHERIUM
Relative of modern hyraxes, it lived 40 million years ago.

INDRICOTHERIUM
The largest land mammal ever lived 30 million years ago.

URSAVUS
Thought to be the ancestor of modern bears, it lived 20 million years ago.

WOOLLY MAMMOTHS
Ice-age mammals related to modern elephants that lived 350–10 thousand years ago.

Monotremes

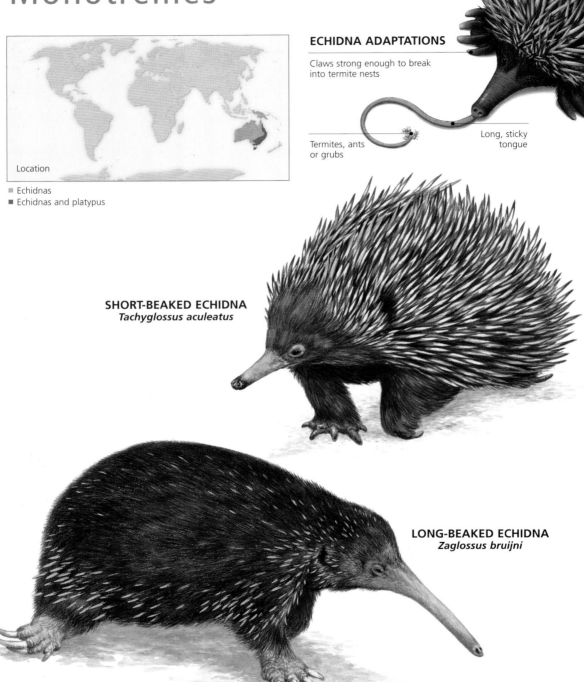

Location

- Echidnas
- Echidnas and platypus

ECHIDNA ADAPTATIONS

Claws strong enough to break into termite nests

Termites, ants or grubs

Long, sticky tongue

SHORT-BEAKED ECHIDNA
Tachyglossus aculeatus

LONG-BEAKED ECHIDNA
Zaglossus bruijni

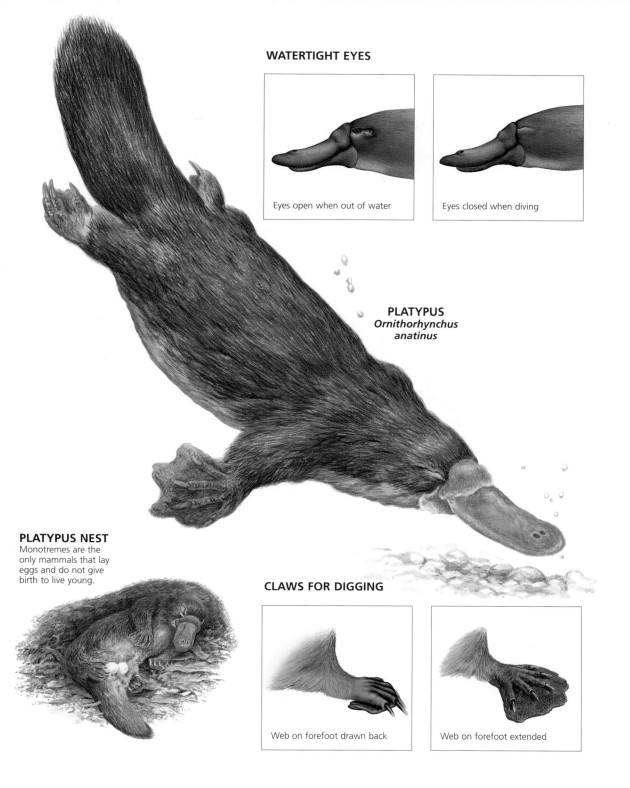

WATERTIGHT EYES

Eyes open when out of water

Eyes closed when diving

PLATYPUS
Ornithorhynchus anatinus

PLATYPUS NEST
Monotremes are the only mammals that lay eggs and do not give birth to live young.

CLAWS FOR DIGGING

Web on forefoot drawn back

Web on forefoot extended

Marsupials

Location

- American marsupials
- Australasian marsupials

NUMBAT
Myrmecobius fasciatus

REPRODUCTIVE ORGANS

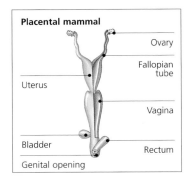

Placental mammal

- Ovary
- Fallopian tube
- Uterus
- Vagina
- Bladder
- Rectum
- Genital opening

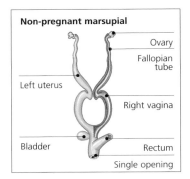

Non-pregnant marsupial

- Ovary
- Fallopian tube
- Left uterus
- Right vagina
- Bladder
- Rectum
- Single opening

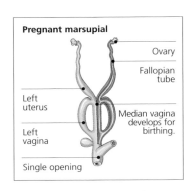

Pregnant marsupial

- Ovary
- Fallopian tube
- Left uterus
- Median vagina develops for birthing.
- Left vagina
- Single opening

STARTING OUT

Embryonic young leaves birth canal.

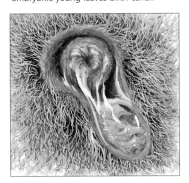

It makes its way to its mother's pouch.

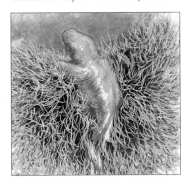

Finds nipple inside the pouch and suckles.

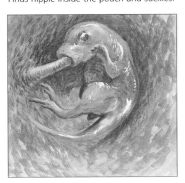

JUMPING INTO THE POUCH

The young kangaroo pushes front feet and head into the pouch.

It rolls forward into the pouch.

Then twists around so it can see and jump out of the pouch.

RED KANGAROO
Macropus rufus

STRIPED POSSUM
Dactylopsila trivirgata

TASMANIAN DEVIL
Sarcophilus laniarius

Marsupials

SPOTTED CUSCUS
Spilocuscus sp.

COMMON WOMBAT
Vombatus ursinus

SQUIRREL GLIDER
Petaurus norfolcensis

WATER OPOSSUM
Chironectes minimus

WOOLLY OPOSSUM
Caluromys derbianus

SPOTTED-TAILED QUOLL
Dasyurus maculatus

KOALA
Phascolarctos cinereus

BILBY
Macrotis lagotis

89

Kangaroos and Wallabies

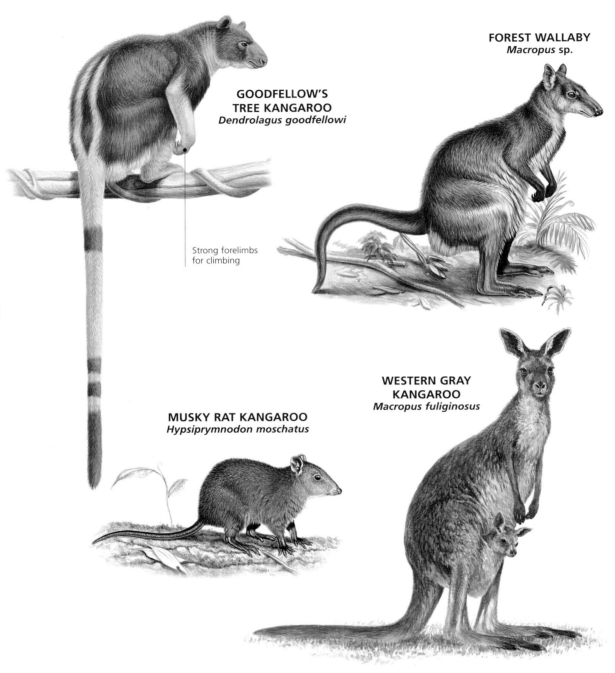

**GOODFELLOW'S
TREE KANGAROO**
Dendrolagus goodfellowi

Strong forelimbs
for climbing

FOREST WALLABY
Macropus sp.

MUSKY RAT KANGAROO
Hypsiprymnodon moschatus

**WESTERN GRAY
KANGAROO**
Macropus fuliginosus

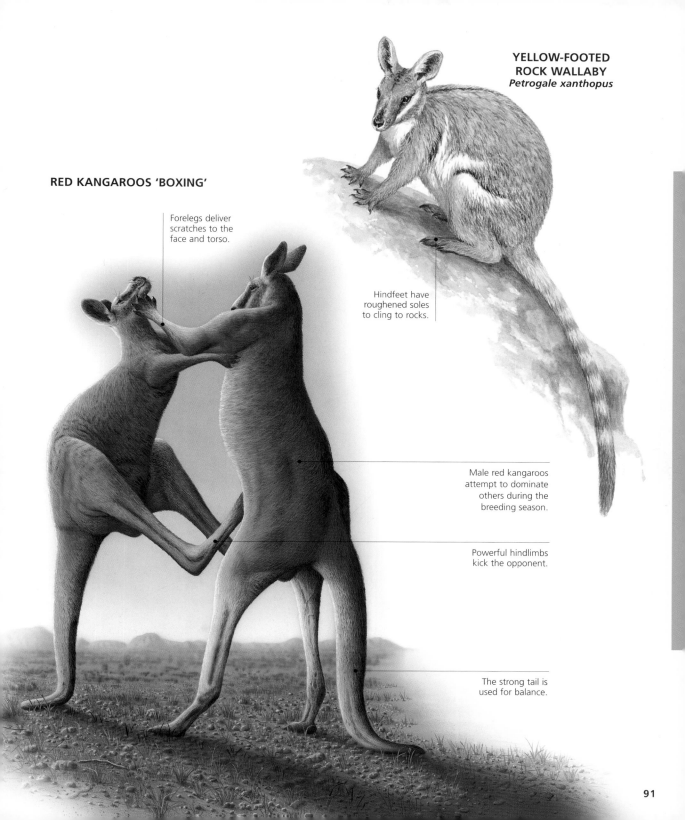

**YELLOW-FOOTED
ROCK WALLABY**
Petrogale xanthopus

RED KANGAROOS 'BOXING'

Forelegs deliver
scratches to the
face and torso.

Hindfeet have
roughened soles
to cling to rocks.

Male red kangaroos
attempt to dominate
others during the
breeding season.

Powerful hindlimbs
kick the opponent.

The strong tail is
used for balance.

Xenarthrans and Pangolins

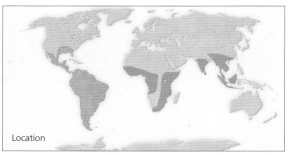

TAMANDUA
Tamandua sp.

MANED THREE-TOED SLOTH
Bradypus torquatus

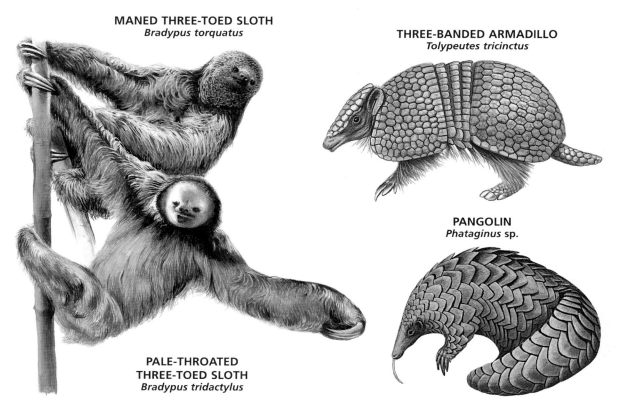

**PALE-THROATED
THREE-TOED SLOTH**
Bradypus tridactylus

THREE-BANDED ARMADILLO
Tolypeutes tricinctus

PANGOLIN
Phataginus sp.

Xenarthrans and Pangolins

MAMMALS

GIANT ANTEATER
Myrmecophaga tridactyla

Baby anteaters are carried on their mother's back.

Long bushy tail is covered with coarse fur.

Tongue can be more than two feet (60 cm) long.

Tubelike snout has no teeth.

It walks on its knuckles to protect claws from wear.

Termites are the anteater's main diet.

Insect-eating Mammals

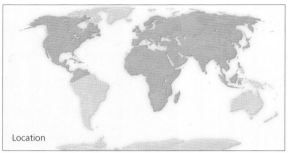

Location

■ Insect-eating mammals

WESTERN EUROPEAN HEDGEHOG
Erinaceus europaeus

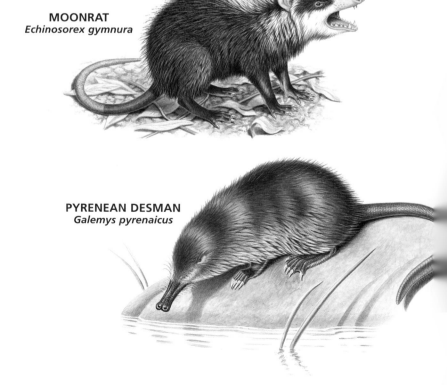

TYPES OF NOSES

Algerian hedgehog
Short pointed snout
with bristles

European mole
Sensitive nose to smell
and feel for prey

Pyrenean desman
Flexible probe to find
insects underwater

MOONRAT
Echinosorex gymnura

PYRENEAN DESMAN
Galemys pyrenaicus

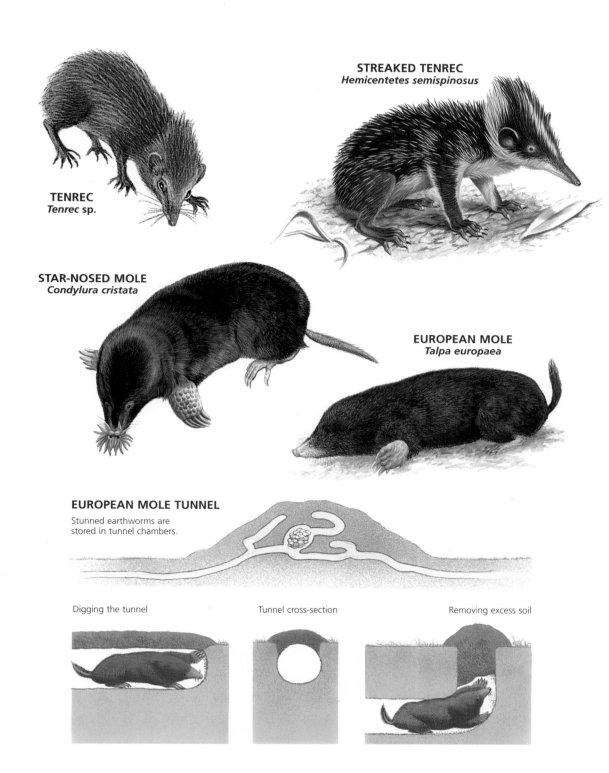

TENREC
Tenrec sp.

STREAKED TENREC
Hemicentetes semispinosus

STAR-NOSED MOLE
Condylura cristata

EUROPEAN MOLE
Talpa europaea

EUROPEAN MOLE TUNNEL
Stunned earthworms are
stored in tunnel chambers.

Digging the tunnel

Tunnel cross-section

Removing excess soil

Flying Lemurs and Tree Shrews

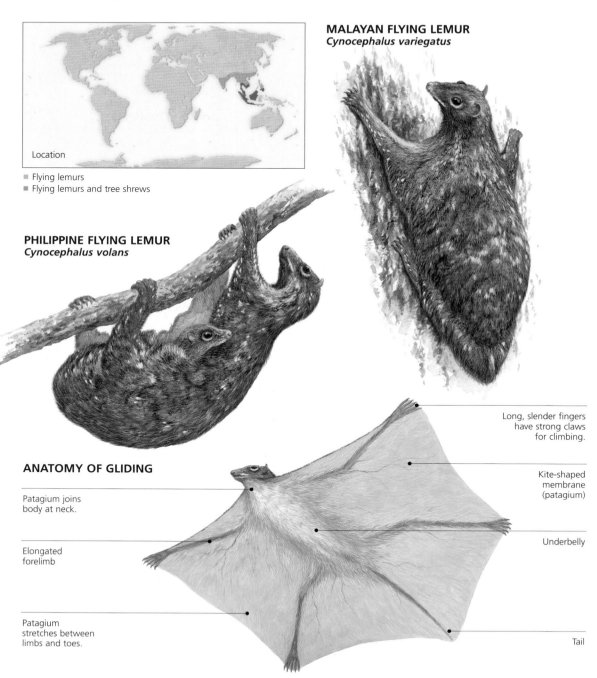

Location

- Flying lemurs
- Flying lemurs and tree shrews

MALAYAN FLYING LEMUR
Cynocephalus variegatus

PHILIPPINE FLYING LEMUR
Cynocephalus volans

ANATOMY OF GLIDING

Patagium joins
body at neck.

Elongated
forelimb

Patagium
stretches between
limbs and toes.

Long, slender fingers
have strong claws
for climbing.

Kite-shaped
membrane
(patagium)

Underbelly

Tail

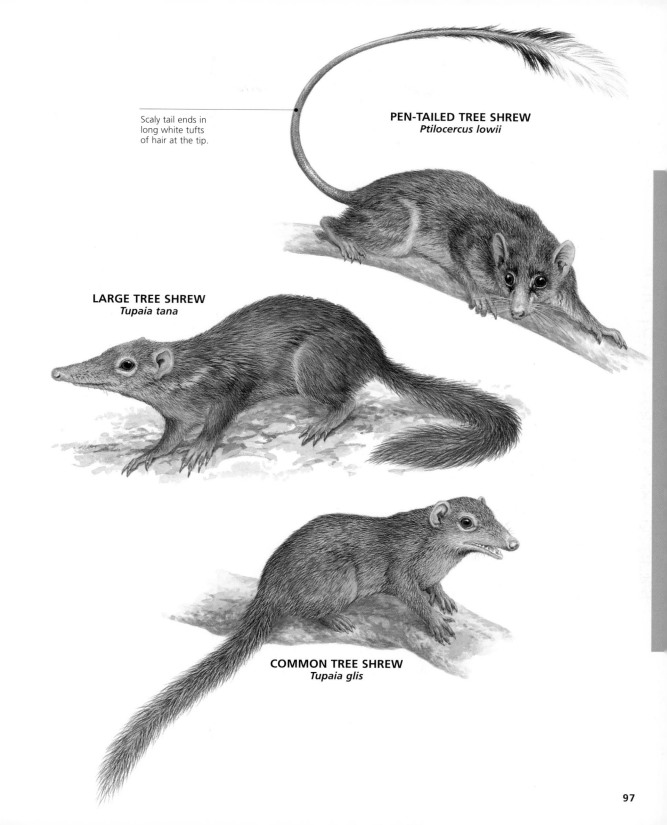

PEN-TAILED TREE SHREW
Ptilocercus lowii

Scaly tail ends in
long white tufts
of hair at the tip.

LARGE TREE SHREW
Tupaia tana

COMMON TREE SHREW
Tupaia glis

Bats

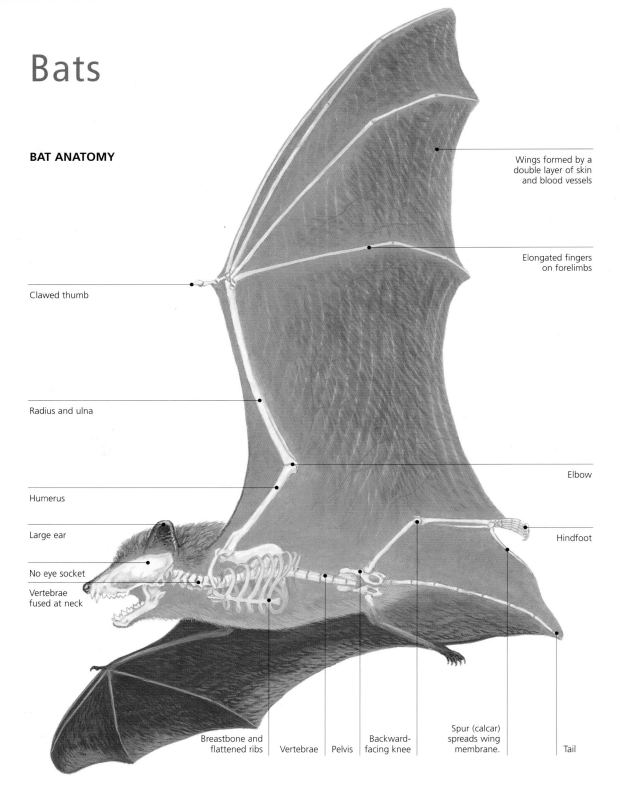

BAT ANATOMY

Wings formed by a
double layer of skin
and blood vessels

Elongated fingers
on forelimbs

Clawed thumb

Radius and ulna

Elbow

Humerus

Large ear

Hindfoot

No eye socket

Vertebrae
fused at neck

Breastbone and
flattened ribs

Vertebrae

Pelvis

Backward-
facing knee

Spur (calcar)
spreads wing
membrane.

Tail

ECHOLOCATION

Insect-eating bats use pulses of ultrasound to detect and capture prey.

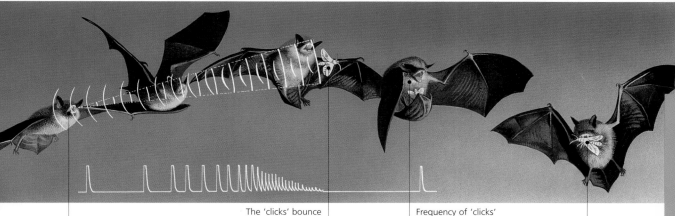

Bat emits rapid 'clicks' of sound.

The 'clicks' bounce off prey back to bat, revealing its location.

Frequency of 'clicks' increases as bat nears to pinpoint prey.

Bat seizes prey.

VARIATIONS

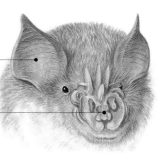

Trident bat

Ear detects reflected sound.

Nose emits 'clicks' of sound.

Whiskered bat

Ear detects reflected sound.

Mouth emits 'clicks' of sound.

EARS AND NOSES

Long-eared bat
Large ears with flaps may improve accuracy of echolocation.

Lesser bare-backed fruit bat
Tube-shaped nostrils sniff out food.

Tent-building bat
Flaps of skin on leaf-shaped nose help detect echoes from prey.

Old World Fruit Bats

Location

■ Old World fruit bats

**GAMBIAN EPAULETTED
FRUIT BAT**
Epomorphorus gambianus

HAMMER-HEADED FRUIT BAT
Hypsignathus monstrosus

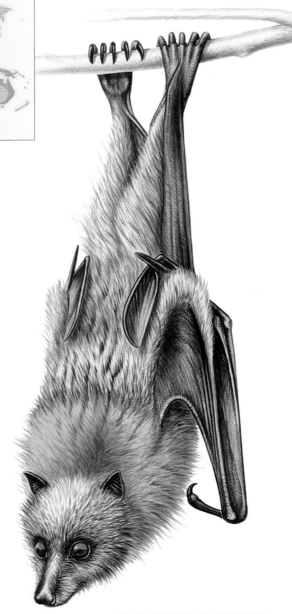

GRAY-HEADED FLYING FOX
Pteropus poliocephalus

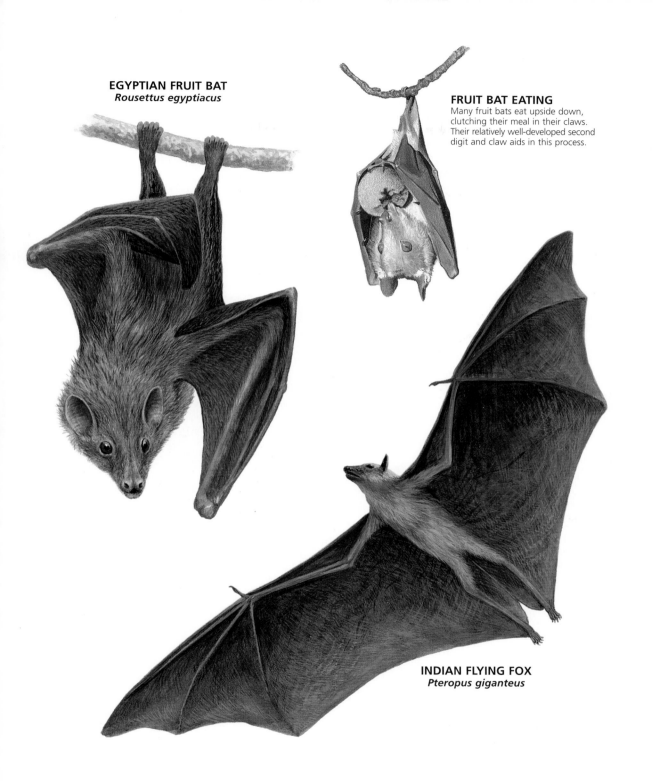

EGYPTIAN FRUIT BAT
Rousettus egyptiacus

FRUIT BAT EATING
Many fruit bats eat upside down, clutching their meal in their claws. Their relatively well-developed second digit and claw aids in this process.

INDIAN FLYING FOX
Pteropus giganteus

Insect-eating Bats

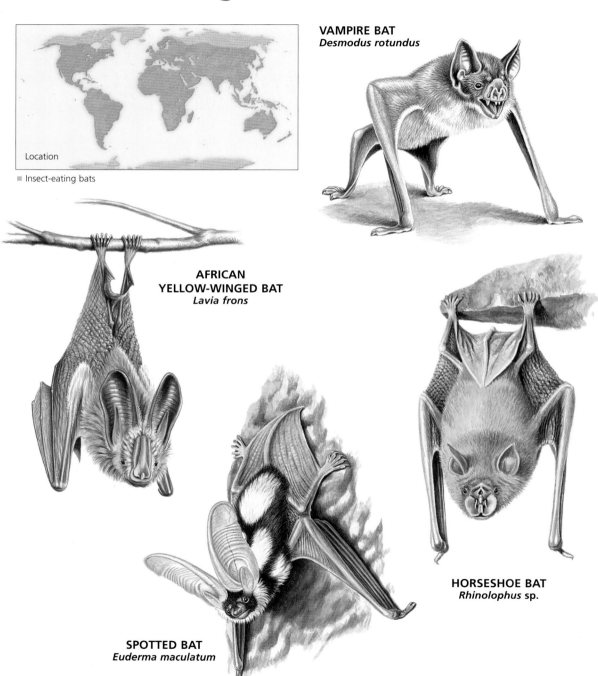

Location

- Insect-eating bats

VAMPIRE BAT
Desmodus rotundus

**AFRICAN
YELLOW-WINGED BAT**
Lavia frons

HORSESHOE BAT
Rhinolophus sp.

SPOTTED BAT
Euderma maculatum

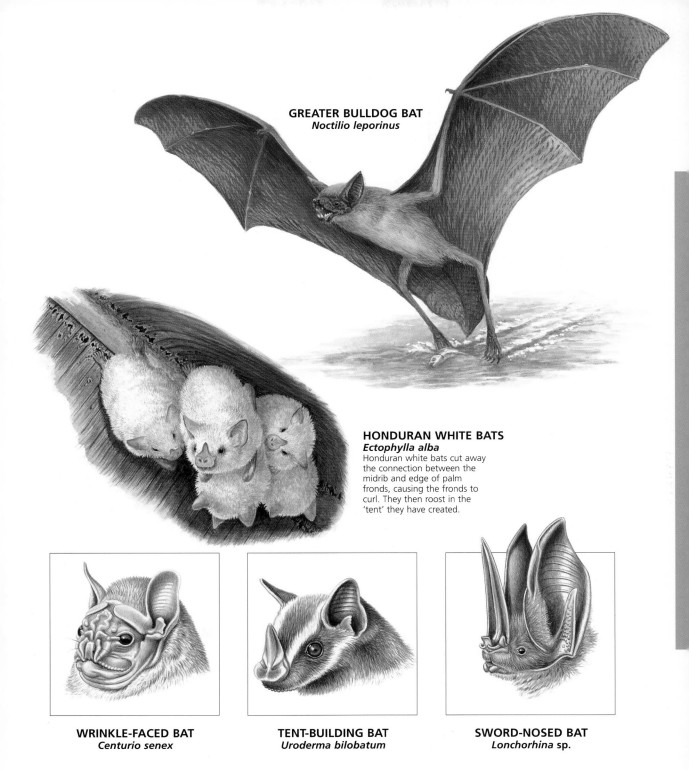

GREATER BULLDOG BAT
Noctilio leporinus

HONDURAN WHITE BATS
Ectophylla alba
Honduran white bats cut away
the connection between the
midrib and edge of palm
fronds, causing the fronds to
curl. They then roost in the
'tent' they have created.

WRINKLE-FACED BAT
Centurio senex

TENT-BUILDING BAT
Uroderma bilobatum

SWORD-NOSED BAT
Lonchorhina sp.

Primates

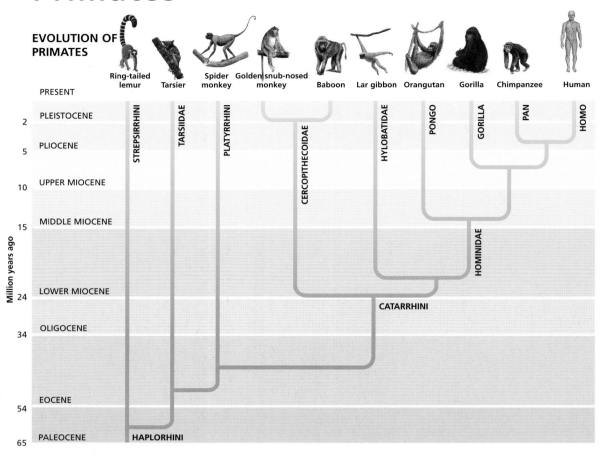

EVOLUTION OF PRIMATES

Ring-tailed lemur | Tarsier | Spider monkey | Golden snub-nosed monkey | Baboon | Lar gibbon | Orangutan | Gorilla | Chimpanzee | Human

PRESENT

Million years ago		
2	PLEISTOCENE	
5	PLIOCENE	
10	UPPER MIOCENE	
15	MIDDLE MIOCENE	
24	LOWER MIOCENE	
34	OLIGOCENE	
54	EOCENE	
65	PALEOCENE	

STREPSIRRHINI
TARSIIDAE
PLATYRRHINI
CERCOPITHECOIDAE
HYLOBATIDAE
PONGO
GORILLA
PAN
HOMO
HOMINIDAE
CATARRHINI
HAPLORHINI

PRIMATE FEET AND HANDS

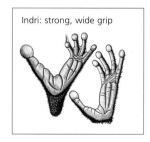

Indri: strong, wide grip

Aye-aye: uses claws to climb

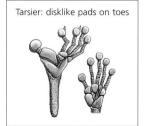

Tarsier: disklike pads on toes

Gorilla: broad to support weight

PROBOSCIS MONKEY
Asian Old World
monkey

LAR GIBBON
Lesser ape

COTTON-TOP TAMARIN
New World monkey

MANDRILL
African Old World monkey

RING-TAILED LEMUR
Lower primate

Lower Primates

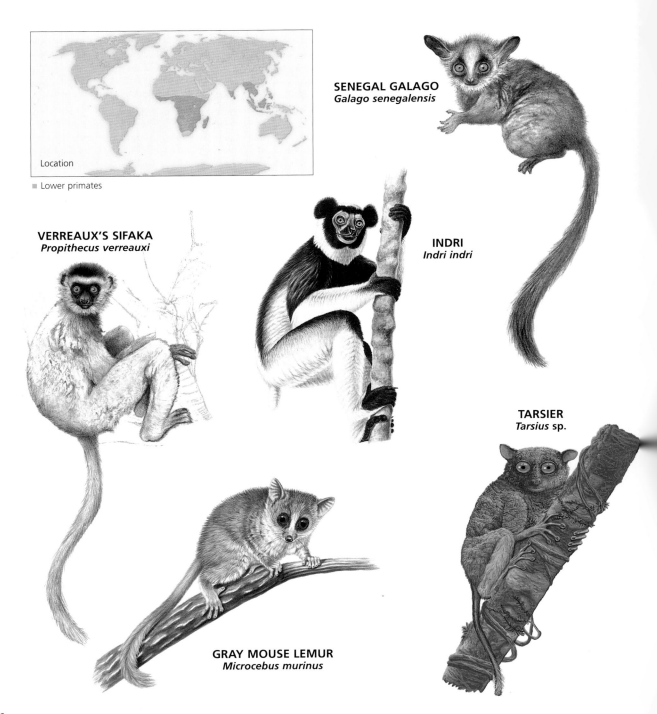

Location

■ Lower primates

SENEGAL GALAGO
Galago senegalensis

VERREAUX'S SIFAKA
Propithecus verreauxi

INDRI
Indri indri

TARSIER
Tarsius sp.

GRAY MOUSE LEMUR
Microcebus murinus

Ring-tailed lemurs live in troops of up to 20 individuals.

Black and white rings on tail

Their diet consists of foraged fruit and insects.

LOWER PRIMATE ADAPTATIONS

An enlarged toilet claw, used for grooming, is only on the second digit.

Troop males use horny spurs on their wrists to mark their territory.

RING-TAILED LEMUR
Lemur catta

DENTAL COMB
Used in mutual grooming

Dental comb on lower jaw

Lower incisors and canines

Lower molar

Lower Primates

FORK-MARKED LEMUR
Phaner furcifer

GOLDEN POTTO
Arctocebus calabarensis

RED RUFFED LEMUR
Varecia variegata rubra

SPECTRAL TARSIER
Tarsius spectrum

DIADEMED SIFAKA
Propithecus diadema

BROWN MOUSE LEMUR
Microcebus rufus

COQUEREL'S MOUSE LEMUR
Microcebus coquereli

PHILIPPINE TARSIER
Tarsius syrichta

SLOW LORIS
Nycticebus coucang

AYE-AYE
Daubentonia madagascariensis

109

Higher Primates

NEW WORLD
Woolly monkey

OLD WORLD
Vervet monkey

Narrow septum
with nostrils that
face downward

Larger species
have a prehensile
tail that acts as
a fifth limb.

Broad septum
with nostrils that
face sideways

Thumbs on forefeet
and hindfeet are highly
opposable (able to
grasp objects easily).

Tail not prehensile

Thumbs are not
highly opposable
to other fingers.

SPIDER MONKEY

Spiders monkeys move through the trees using their extended limbs or swinging from their prehensile tail.

HIGHER PRIMATE FACES

Greater white-nosed monkey

Japanese macaque

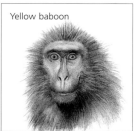

Yellow baboon

Tonkin snub-nosed monkey

LAR GIBBON

Using their bodies as a pendulum, lar gibbons swing hand-over-hand in an energy-saving process called brachiation. They can travel up to 10 feet (3 m) each swing.

Ball-and-socket joints in the wrists increase mobility.

African Old World Monkeys

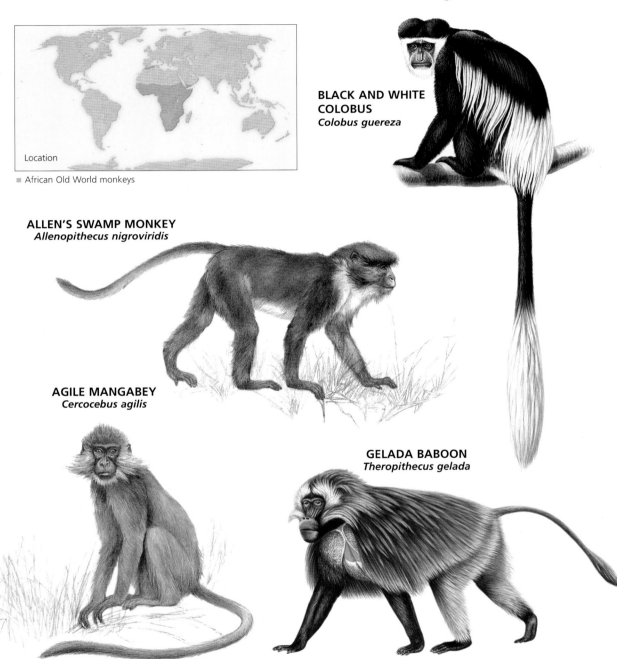

Location
■ African Old World monkeys

BLACK AND WHITE COLOBUS
Colobus guereza

ALLEN'S SWAMP MONKEY
Allenopithecus nigroviridis

AGILE MANGABEY
Cercocebus agilis

GELADA BABOON
Theropithecus gelada

DIANA MONKEY
Cercopithecus diana

HAMADRYAS BABOONS
Papio hamadryas

VERVET MONKEY
Chlorocebus aethiops

DRILL
Mandrillus leucophaeus

BLACK COLOBUS
Colobus satanas

113

Asian Old World Monkeys

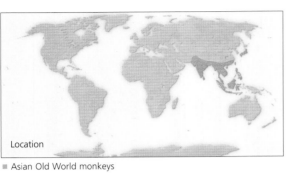

Location

■ Asian Old World monkeys

DOUC LANGUR
Pygathrix nemaeus

PROBOSCIS MONKEY
Nasalis larvatus

**CELEBES CRESTED
MACAQUE**
Macaca nigra

GOLDEN SNUB-NOSED MONKEY
Pygathrix roxellana

PIGTAIL MACAQUE
Macaca nemestrina

STUMP-TAILED MACAQUE
Macaca arctoides

DUSKY LEAF MONKEY
Trachypithecus obscurus

115

New World Monkeys

Location

■ New World monkeys

MIDAS TAMARIN
Saguinus midas

BROWN CAPUCHIN
Cebus apella

**NORTHERN NIGHT
MONKEY**
Aotus trivirgatus

RED HOWLER MONKEY
Alouatta seniculus

MURIQUI
Brachyteles arachnoides

COTTON-TOP TAMARINS
Saguinus oedipus
Cotton-top tamarins live in family troops of 10–12 individuals with a single breeding pair. Older siblings and troop males help raise and carry the pair's offspring.

RED UAKARI
Cacajao calvus

MASKED TITI
Callicebus personatus

Great Apes

Location
■ Apes

MOUNTAIN GORILLA
Gorilla gorilla beringei

WESTERN LOWLAND GORILLA
Gorilla gorilla gorilla

ORANGUTAN
Pongo pygmaeus

Head of male orangutan

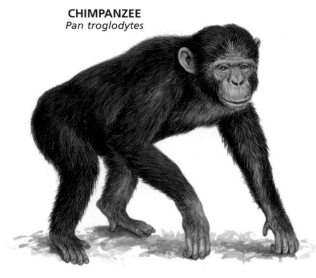

CHIMPANZEE
Pan troglodytes

BONOBO
Pan paniscus

119

Gorillas

SKULL COMPARISON

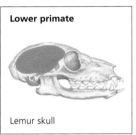

Lower primate

Lemur skull

Higher primate

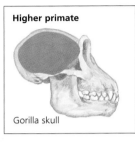

Gorilla skull

Human skull

FAMILY GROUPS

Mountain gorillas live in closely knit family groups of related females in the mountain forests of Africa.

A large and dominant silverback male leads and protects the group.

Gorillas build nests in the trees each night for warmth and safety.

All apes have a fully opposable thumb that is used to grasp objects.

Gorillas are plant eaters (herbivores) and spend most of the day eating.

121

Carnivores

EVOLUTION OF CARNIVORES

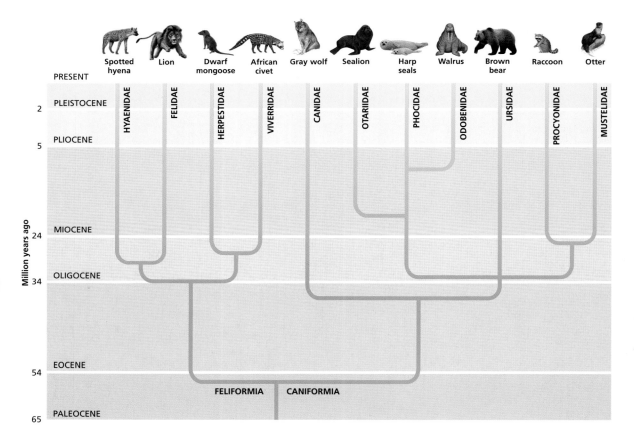

DENTAL ADAPTATIONS
■ Slicing teeth ■ Grinding teeth

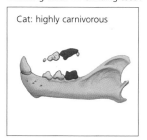

Cat: highly carnivorous

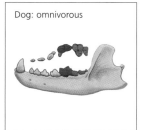

Dog: omnivorous

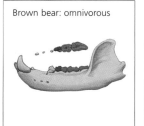

Brown bear: omnivorous

Giant panda: herbivorous

TYPES OF CARNIVORES

Ocelot
Family *Felidae*

Pharoah hound
Family *Canidae*

Brown bear
Family *Ursidae*

Raccoon
Family *Procyonidae*

Weasel
Family *Mustelidae*

African civet
Family *Viverridae*

Spotted hyena
Family *Hyaenidae*

Pounces
silently

Grabs prey
by throat

Prey unable
to breathe

Tiger kills prey
by suffocation.

Tiger tracks prey
while camouflaged
by long grass.

TIGER HUNTING

Wolves, Dogs and Foxes

GRAY WOLF PACK

Wolves are efficient hunters who work in a group to exhaust their prey.

Wolf cubs stay with their birth pack until they mature physically.

Wolves communicate with howls, body language and facial expressions.

Wolf packs enforce a strict social hierarchy within the group.

Gray wolf

Female red fox with pups

African wild dogs fighting

WOLF TAILS

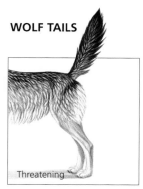

Threatening

Relaxed

Fearful

Submissive

Ready to attack

DOMESTIC DOG

DOMESTIC DOG SKELETON

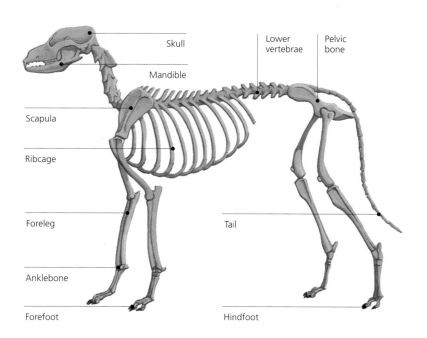

Skull

Lower vertebrae

Pelvic bone

Mandible

Scapula

Ribcage

Foreleg

Tail

Anklebone

Forefoot

Hindfoot

Wolves, Dogs and Foxes

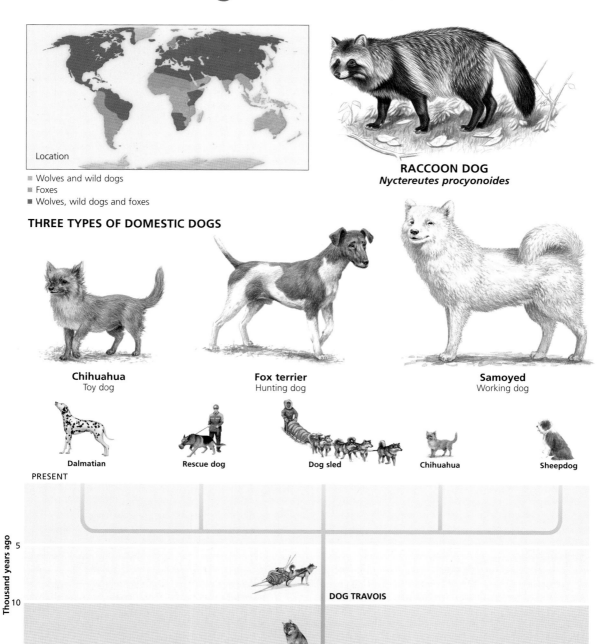

Location

- Wolves and wild dogs
- Foxes
- Wolves, wild dogs and foxes

RACCOON DOG
Nyctereutes procyonoides

THREE TYPES OF DOMESTIC DOGS

Chihuahua
Toy dog

Fox terrier
Hunting dog

Samoyed
Working dog

Dalmatian

Rescue dog

Dog sled

Chihuahua

Sheepdog

PRESENT

Thousand years ago

5

10

15

DOG TRAVOIS

ANCESTRAL WOLF

MANED WOLF
Chrysocyon brachyurus

GRAY WOLF
Canis lupus

AFRICAN WILD DOG
Lycaon pictus

BUSH DOG
Speothos venaticus

DINGO
Canis lupus dingo

RED WOLF
Canis rufus

127

Wolves, Dogs and Foxes

ANATOMY OF HUNTING

Pricked-up ears
to pinpoint sound

Forward-facing eyes
to perceive depth

Sensitive nose
to detect prey

Teeth with shearing
edges to slice flesh

BAT-EARED FOX
Otocyon megalotis

GOLDEN JACKAL
Canis aureus

ARCTIC FOX
Alopex lagopus

Arctic foxes have
relatively small ears
and a thick pelt to
prevent heat loss.

Carnivores

MAMMALS

128

COYOTE
Canis latrans

GRAY FOX
Urocyon cinereoargenteus

KIT FOX
Vulpes macrotis

COYOTE EXPRESSIONS

Submissive

Combative

Defensive

Friendly

Playful

Bears

EVOLUTION OF BEARS

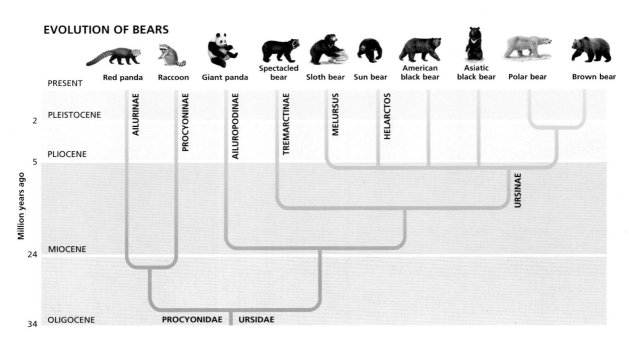

| | Red panda | Raccoon | Giant panda | Spectacled bear | Sloth bear | Sun bear | American black bear | Asiatic black bear | Polar bear | Brown bear |

PRESENT

PLEISTOCENE

2

PLIOCENE

5

Million years ago

MIOCENE

24

OLIGOCENE

34

AILURINAE
PROCYONINAE
AILUROPODINAE
TREMARCTINAE
MELURSUS
HELARCTOS
URSINAE

PROCYONIDAE   URSIDAE

BEAR SIZES

Sun bear

Giant panda

Asiatic black bear

Sloth bear

American black bear

Spectacled bear

Brown bear

Polar bear

FOOT COMPARISON

American black bear

Long, sharp claws on forefeet

Walks flat on hindfeet

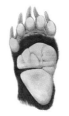

Giant panda

Short, blunt claws on forefeet

Walks on toes of hindfeet

FOREFOOT COMPARISON

Brown bear
Long claws for digging

Giant panda
Modified wrist acts like digit

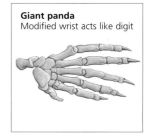

TREE CLIMBERS
American black bears are agile climbers and often snooze in the safety of the treetops.

BROWN BEAR

BROWN BEAR SKELETON

Skull

Mandible

Scapula

Ribcage

Pelvic bone

Hip

Knee

Foreleg

Claws

Anklebone

Eurasian and New World Bears

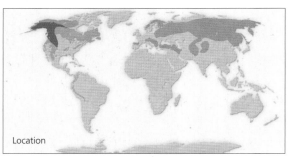

Location

- ■ Eurasian bear (brown bear)
- ■ New World bears
- ■ Eurasian bear and New World bears

AMERICAN BLACK BEAR REPRODUCTION

Winter
Hibernation and birth of young

Spring
Nursing and rearing young

Summer
Mating and intense feeding

Autumn
Delayed implantation of embryo

BROWN BEAR
Ursus arctos

SPECTACLED BEAR
Tremarctos ornatus

**AMERICAN
BLACK BEAR**
Ursus americanus

133

Polar Bear

Location

■ Polar bear

POLAR BEAR CUBS

POLAR BEAR YEAR

1. Late February–April
Mother and pups leave den and mother replenishes fat stores.

2. March–April
Adult males travel in search of breeding females, then fight for breeding rights.

8. December–February
Except for pregnant females, polar bears actively feed throughout winter.

7. November–January
Pregnant female gives birth and nurses cubs in the den.

POLAR BEAR
Ursus maritimus

3. April–May
Mating occurs over two weeks and eggs are fertilized, but are not implanted until September.

4. May–June
At two-and-a-half years of age, young bears are weaned from their mothers.

6. August–late November
Sea ice melts and polar bears come ashore, living off their fat reserves until the sea freezes again.

5. April–July
Ringed seal pups, born in April, are easy prey for polar bears.

Asian Bears

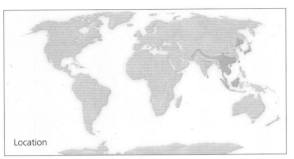

Location

- Asian bears
- Giant panda

SLOTH BEAR
Melursus ursinus

ASIATIC BLACK BEAR
Ursus thibetanus

SUN BEAR
Reaching only 3 feet
(1 m) in height, the
sun bear is the
world's smallest.

SUN BEAR
Helarctos malayanus

GIANT PANDA
Ailuropoda melanoleuca

Mustelids

Location

■ Mustelids (badgers, weasels, otters, skunks)

AMERICAN BADGER
Taxidea taxus

EUROPEAN BADGER SETT (BURROW)

While badgers are usually solitary creatures, they build an elaborate system of communal tunnels, called a sett, with individual nesting chambers for each badger.

Entrance

Individual chamber

Individual chamber

EUROPEAN POLECAT
Mustela putorius

MARBLED POLECAT
Vormela peregusna

EURASIAN BADGER
Meles meles

HOG BADGER
Arctonyx collaris

MOLINA'S
HOG-NOSED SKUNK
Conepatus chinga

HUMBOLDT'S
HOG-NOSED SKUNK
Conepatus humboldtii

SKUNK DEFENSE
Mustelids have musk glands
that are used for defense
and marking territory.

Confronts predator face-to-face

Raises tail in warning

Rises up on forelegs

Ejects foul-smelling liquid

Mustelids

SEA OTTER
Sea otters float on their back and use small stones to crack open sea urchins.

MARTEN
Martens are agile hunters with powerful forelimbs and a long tail. They chase prey by leaping through the trees.

EUROPEAN OTTER
Lutra lutra

GIANT OTTER
Pteronura brasiliensis

ERMINE
Mustela erminea
Summer coat

AFRICAN STRIPED WEASEL
Poecilogale albinucha

ERMINE
Winter coat

LEAST WEASEL
Mustela nivalis

EUROPEAN MINK
Mustela lutreola

YELLOW-THROATED MARTEN
Martes flavigula

EUROPEAN PINE MARTEN
Martes martes

Catlike Carnivores

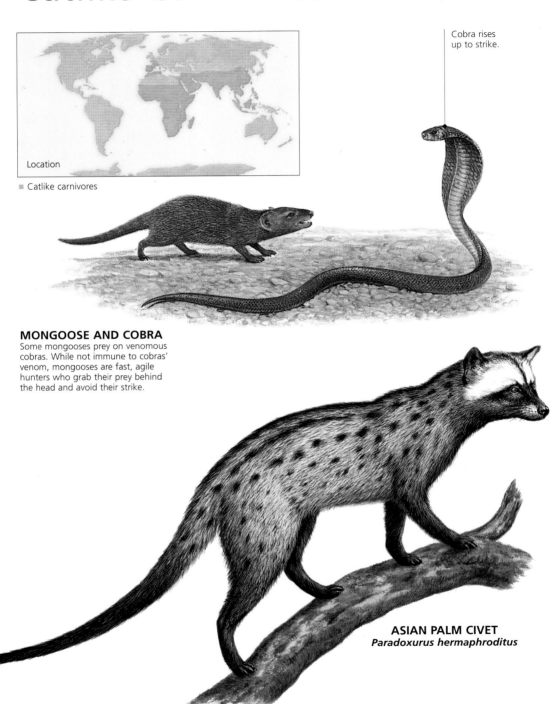

Location

■ Catlike carnivores

Cobra rises
up to strike.

MONGOOSE AND COBRA
Some mongooses prey on venomous
cobras. While not immune to cobras'
venom, mongooses are fast, agile
hunters who grab their prey behind
the head and avoid their strike.

ASIAN PALM CIVET
Paradoxurus hermaphroditus

MEERKATS
Suricata suricatta

Meerkats live in family groups, complete with sentinels to protect the group from predators.

FOSSA
Cryptoprocta ferox

BANDED LINSANG
Prionodon linsang

AFRICAN CIVET
Civettictis civetta

Raccoons and Hyenas

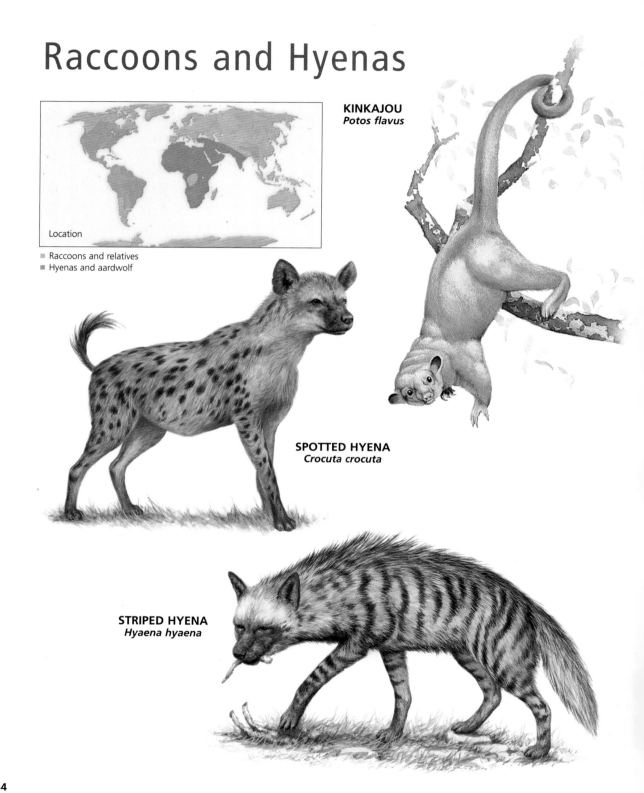

Location

- Raccoons and relatives
- Hyenas and aardwolf

KINKAJOU
Potos flavus

SPOTTED HYENA
Crocuta crocuta

STRIPED HYENA
Hyaena hyaena

RED PANDA
Ailurus fulgens

SOUTH AMERICAN COATI
Nausa nausa

AARDWOLF
Proteles cristatus

RINGTAIL
Bassariscus astutus

NORTHERN RACCOON
Procyon lotor

Hyena Hunting Tactics

PACK HUNTING

The largest members of the hyena family, spotted hyenas are not exclusively scavengers, but are efficient pack hunters.

They are able to accelerate to speeds of over 30 miles per hour (48 km/h), and will chase their prey at this speed for more than a mile.

Other hyenas nearby join in the chase, until the gemsbok is exhausted. They work together to bring down their prey.

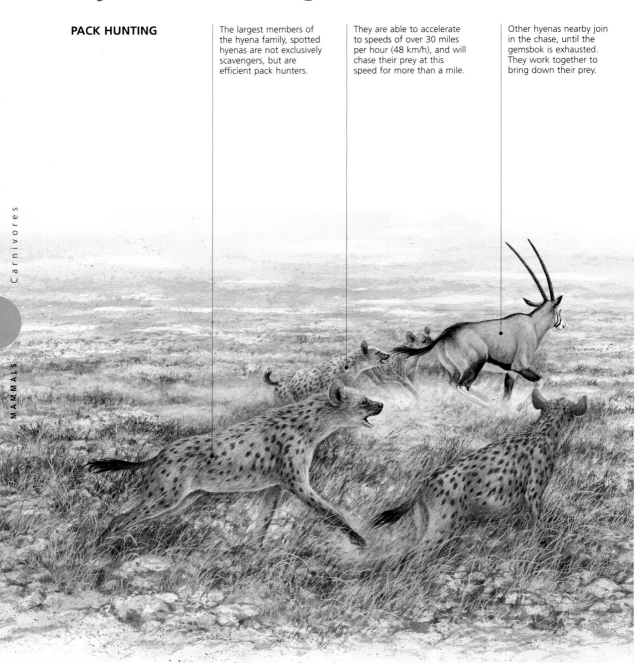

The spotted hyena's jaws are strong enough to crunch through bone; the hydrochloric acid in its stomach breaks down bone.

Rapid eaters, a pack of hyenas can demolish an adult gemsbok in as little as half an hour, this speed necessitated by threats from scavengers.

Despite their reputation as fearsome hunters, lions are just as likely to scavenge a kill from hyenas, who are then forced to abandon it.

STEALING THE KILL

Cats

EVOLUTION OF CATS

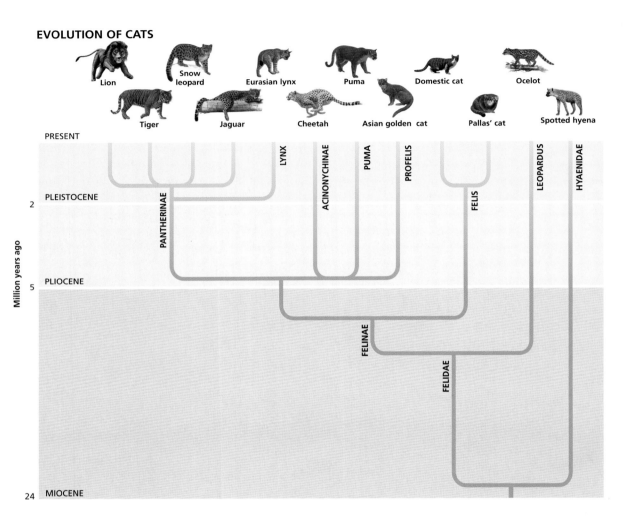

Lion

Snow leopard

Eurasian lynx

Puma

Domestic cat

Ocelot

Tiger

Jaguar

Cheetah

Asian golden cat

Pallas' cat

Spotted hyena

Million years ago

PRESENT

PLEISTOCENE

2

PLIOCENE

5

MIOCENE

24

PANTHERINAE

LYNX

ACINONYCHINAE

PUMA

PROFELIS

FELIS

LEOPARDUS

HYAENIDAE

FELINAE

FELIDAE

MARGAY EXPRESSIONS

Aggressive

Defensive

SPRINGLIKE CLAWS

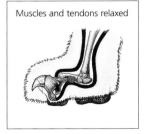

Muscles and tendons relaxed

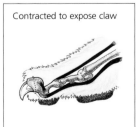

Contracted to expose claw

WILD CAT

WILD CAT SKELETON

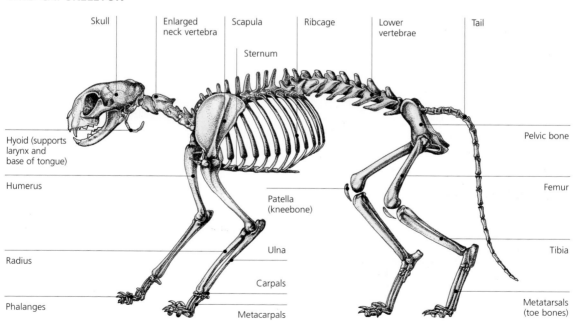

Skull

Enlarged
neck vertebra

Scapula

Sternum

Ribcage

Lower
vertebrae

Tail

Hyoid (supports
larynx and
base of tongue)

Humerus

Radius

Phalanges

Patella
(kneebone)

Ulna

Carpals

Metacarpals

Pelvic bone

Femur

Tibia

Metatarsals
(toe bones)

Great Cats

Location

■ Great cats

OCELOT
Leopardus pardalis

LION PRIDE

There are usually one, but often two related dominant males in the pride.

Lionesses are the primary hunters of the pride, and hunt together to provide meat for the group.

Females breed from the age of four, and mate every 20 minutes for five days when on heat.

Up to three generations of females can live in one pride, which can hold as many as 30 individuals.

Lion cubs nurse for up to six months, before weaning. Cubs are the last to feed from kills.

PUMA
Puma concolor

LION AND LIONESS
Panthera leo

TIGER
Panthera tigris

Great Cats

CLOUDED LEOPARD
Neofelis nebulosa

MARBLED CAT
Pardofelis marmorata

SNOW LEOPARD
Uncia uncia

LEOPARD AND GIRAFFE
Leopards often rest in trees, and will stow a kill on a branch to keep it out of reach of competitors and scavengers.

JAGUAR
Panthera onca

BLACK PANTHER
Panthera pardus
Also known as melanic leopards, panthers have a genetic mutation that raises the amount of melanin (pigment) in the skin and fur.

LEOPARD
Panthera pardus

Lynxes and Cheetah

Location
- Lynxes, caracal and bobcat
- Lynxes, caracal, bobcat and cheetah

EURASIAN LYNX
Lynx lynx

CANADA LYNX
Lynx canadensis

BOBCAT
Lynx rufous

CARACAL
Caracal caracal

SPANISH LYNX
Lynx pardinus

CHEETAH
Acinonyx jubatus

FASTEST ON LAND
Cheetahs can run at speeds of 70 miles
(110 km) per hour in short bursts of speed.

Small Cats

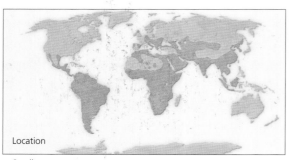

Location

■ Small cats

KODKOD
*Oncifelis
guigna*

LEOPARD CAT
*Prionailurus
bengalensis*

JUNGLE CAT
Felis chaus

FISHING CAT
*Prionailurus
viverrinus*

ANDEAN MOUNTAIN CAT
Felis jacobita

SERVAL
Leptailurus serval

ASIAN GOLDEN CAT
Profelis temmincki

AFRICAN GOLDEN CAT
Profelis aurata

JAGUARUNDI
Herpailurus yagouaroundi

MARGAY
Felis wiedii

Seals, Sealions and Walrus

FUR SEAL

Layer of blubber and thick fur to protect from cold

SEALION SKULL

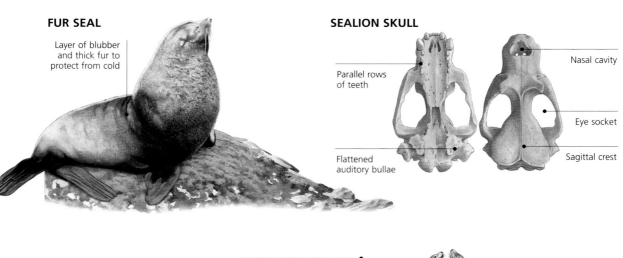

Parallel rows of teeth

Flattened auditory bullae

Nasal cavity

Eye socket

Sagittal crest

AUSTRALIAN SEALION COLONY

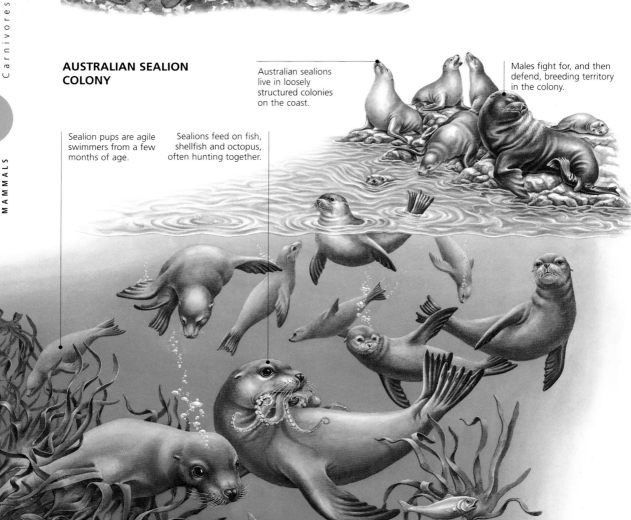

Australian sealions live in loosely structured colonies on the coast.

Males fight for, and then defend, breeding territory in the colony.

Sealion pups are agile swimmers from a few months of age.

Sealions feed on fish, shellfish and octopus, often hunting together.

WALRUS

WALRUS SKELETON

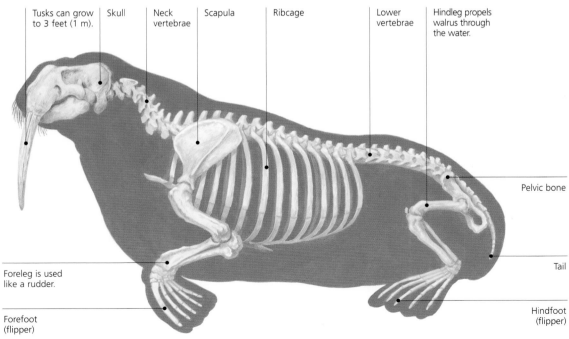

Tusks can grow to 3 feet (1 m).

Skull

Neck vertebrae

Scapula

Ribcage

Lower vertebrae

Hindleg propels walrus through the water.

Pelvic bone

Foreleg is used like a rudder.

Tail

Forefoot (flipper)

Hindfoot (flipper)

Seals, Sealions and Walrus

Location

- Seals and sealions
- Seals, sealions and walrus

BEARDED SEAL
Erignathus barbatus

HARP SEAL
Phoca groenlandica

STELLER SEALION
Eumetopias jubatus

NEW ZEALAND FUR SEAL
Arctocephalus forsteri

BAIKAL SEAL
Phoca sibirica

RIBBON SEAL
Phoca fasciata

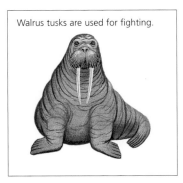
Harbor seal with young

Bull elephant seal guarding territory

Leopard seal

Walrus tusks are used for fighting.

Fur seal showing hind flippers

Hooded seal with inflatable skin pouch

Harp seal pups

Crabeater seals do not eat crabs.

Weddell seals diving off pack ice

Whales, Dolphins and Porpoises

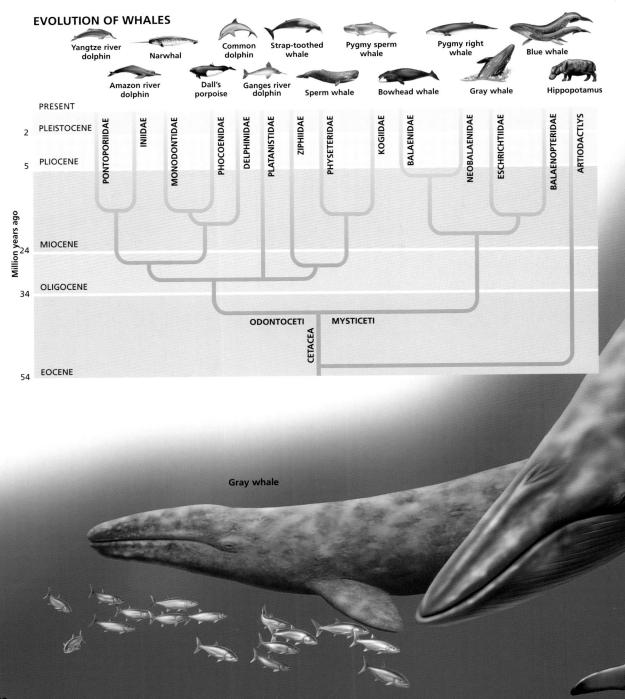

EVOLUTION OF WHALES

Yangtze river dolphin

Narwhal

Common dolphin

Strap-toothed whale

Pygmy sperm whale

Pygmy right whale

Blue whale

Amazon river dolphin

Dall's porpoise

Ganges river dolphin

Sperm whale

Bowhead whale

Gray whale

Hippopotamus

Million years ago

PRESENT

PLEISTOCENE — 2

PLIOCENE — 5

MIOCENE — 24

OLIGOCENE — 34

EOCENE — 54

PONTOPORIIDAE

INIIDAE

MONODONTIDAE

PHOCOENIDAE

DELPHINIDAE

PLATANISTIDAE

ZIPHIIDAE

PHYSETERIDAE

KOGIIDAE

BALAENIDAE

NEOBALAENIDAE

ESCHRICHTIIDAE

BALAENOPTERIDAE

ARTIODACTLYS

ODONTOCETI

MYSTICETI

CETACEA

Gray whale

TYPES OF WHALES
The sei whale and the gray whale
are both baleen whales, while the
killer whale is a dolphin, which is
a type of toothed whale.

Sei whale

Killer whale

WHALE TAILS

Common dolphin

Southern right whale

Blue whale

Humpback whale

Gray whale

Killer whale

Whale Anatomy

Sperm whale head

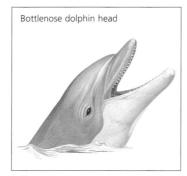

Bottlenose dolphin head

TOOTHED WHALE
Age is measured by counting
rings inside the teeth.

TOOTHED WHALE
Atlantic white-sided dolphin

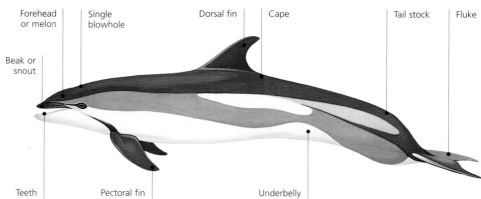

Forehead
or melon

Single
blowhole

Dorsal fin

Cape

Tail stock

Fluke

Beak or
snout

Teeth

Pectoral fin

Underbelly

TOOTHED WHALE SKELETON

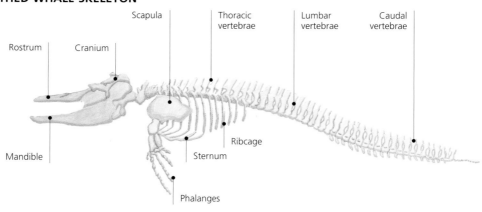

Scapula

Thoracic
vertebrae

Lumbar
vertebrae

Caudal
vertebrae

Rostrum

Cranium

Mandible

Phalanges

Sternum

Ribcage

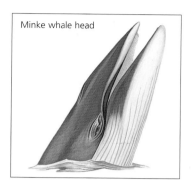

Minke whale head

Southern right whale head

BALEEN WHALE
Age is measured by counting rings inside the waxy ear plug.

BALEEN WHALE
Pygmy right whale

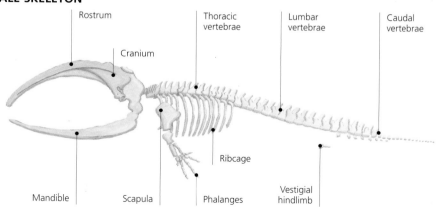

Baleen

Double blowhole

Dorsal fin

Keel

Tail stock

Fluke

Rostrum

Mandible

Pectoral fin

Underbelly

BALEEN WHALE SKELETON

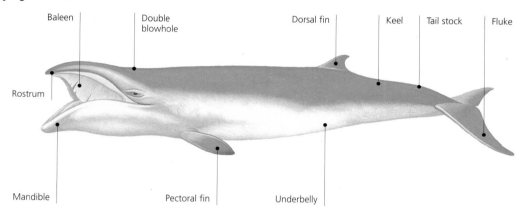

Rostrum

Cranium

Thoracic vertebrae

Lumbar vertebrae

Caudal vertebrae

Mandible

Scapula

Phalanges

Ribcage

Vestigial hindlimb

Baleen Whales

BUBBLENETTING

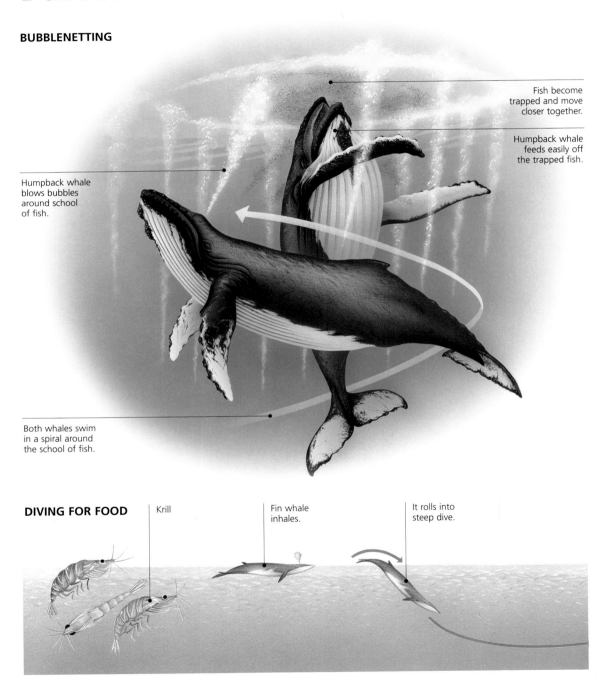

Fish become trapped and move closer together.

Humpback whale feeds easily off the trapped fish.

Humpback whale blows bubbles around school of fish.

Both whales swim in a spiral around the school of fish.

DIVING FOR FOOD

Krill

Fin whale inhales.

It rolls into steep dive.

BALEEN WHALE BEHAVIOR

Lobtailing

Pecslapping

Spyhopping

Breaching

MINKE WHALE
Although many minke whales head to warmer waters for winter breeding, others stay behind. These whales are usually not ready to breed, so save their energy for the following year.

BALEEN WHALE SPOUTS

Humpback whale

Sei whale

Right whale

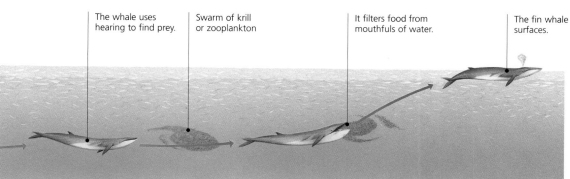

The whale uses hearing to find prey.

Swarm of krill or zooplankton

It filters food from mouthfuls of water.

The fin whale surfaces.

Baleen Whales

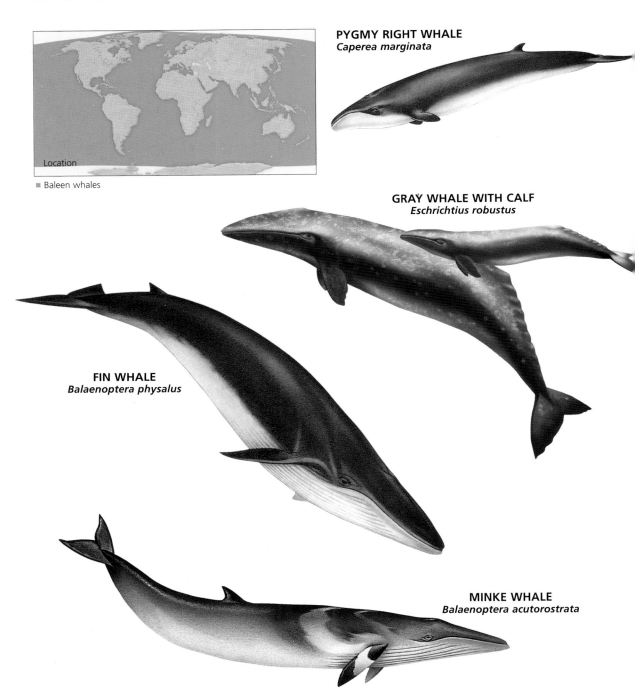

Location

■ Baleen whales

PYGMY RIGHT WHALE
Caperea marginata

GRAY WHALE WITH CALF
Eschrichtius robustus

FIN WHALE
Balaenoptera physalus

MINKE WHALE
Balaenoptera acutorostrata

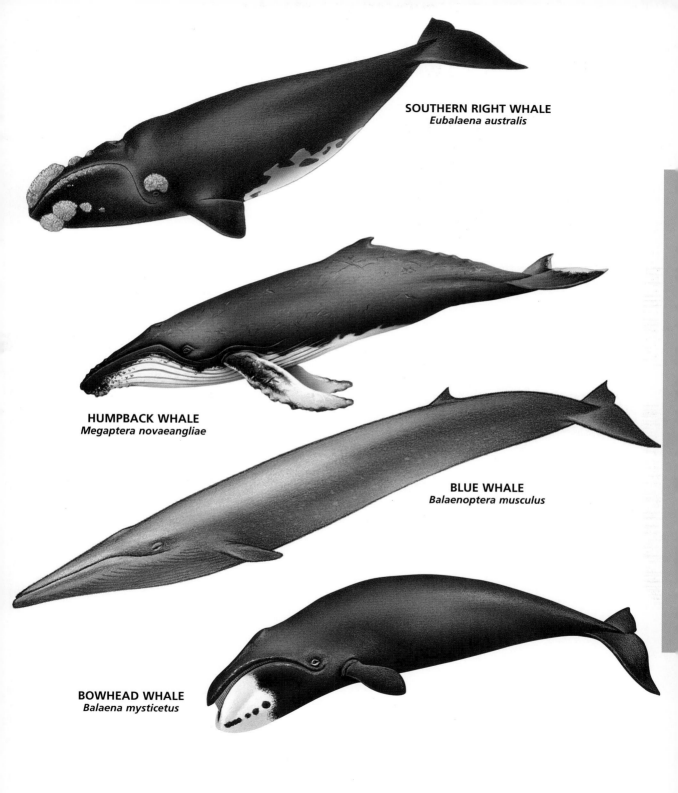

SOUTHERN RIGHT WHALE
Eubalaena australis

HUMPBACK WHALE
Megaptera novaeangliae

BLUE WHALE
Balaenoptera musculus

BOWHEAD WHALE
Balaena mysticetus

Toothed Whales

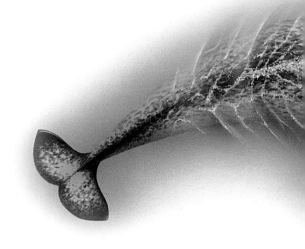

DIVE DEPTHS

Some whales are able to dive to deeper depths than others in search of food. Their bodies are adapted to the intense pressure and cold of deep waters and they can hold their breath for longer periods of time.

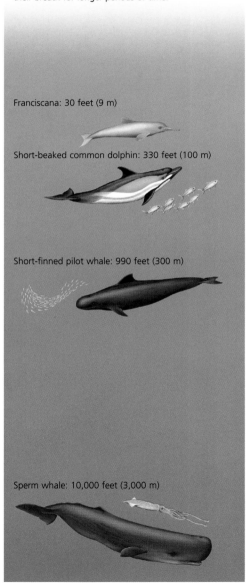

Franciscana: 30 feet (9 m)

Short-beaked common dolphin: 330 feet (100 m)

Short-finned pilot whale: 990 feet (300 m)

Sperm whale: 10,000 feet (3,000 m)

MAKING MUSIC

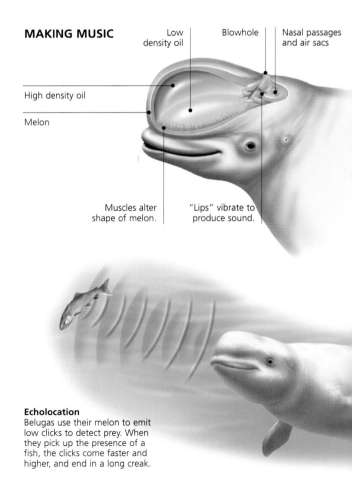

Low density oil

Blowhole

Nasal passages and air sacs

High density oil

Melon

Muscles alter shape of melon.

"Lips" vibrate to produce sound.

Echolocation
Belugas use their melon to emit low clicks to detect prey. When they pick up the presence of a fish, the clicks come faster and higher, and end in a long creak.

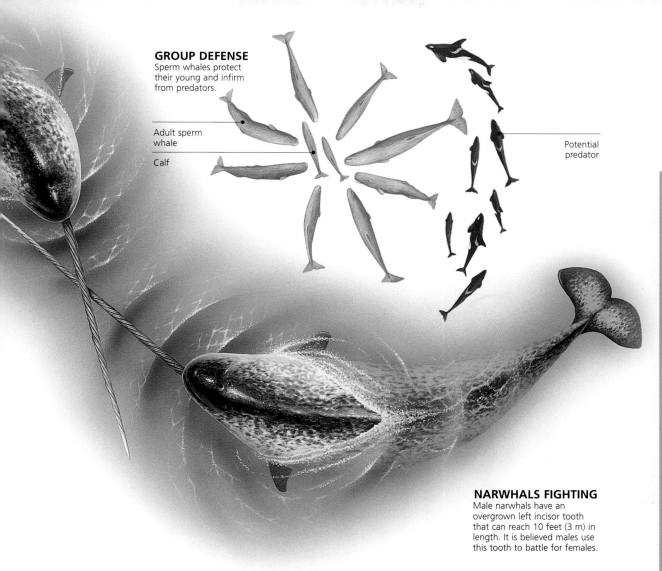

GROUP DEFENSE
Sperm whales protect their young and infirm from predators.

Adult sperm whale

Calf

Potential predator

NARWHALS FIGHTING
Male narwhals have an overgrown left incisor tooth that can reach 10 feet (3 m) in length. It is believed males use this tooth to battle for females.

Types of teeth

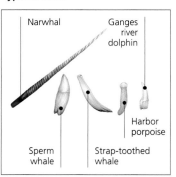

Narwhal

Ganges river dolphin

Sperm whale

Strap-toothed whale

Harbor porpoise

Toothed whale behavior

Touching beaks

Touching pectoral fins

Toothed Whales

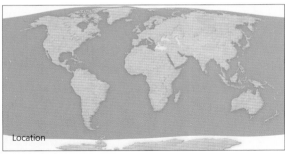

Location

■ Toothed whales

STRAP-TOOTHED WHALE
Mesoplodon layardii

SPERM WHALE DIVING | Sperm whale inhales and prepares to dive. | It dives vertically to about 1,300 feet (394 m), but some may descend to 10,000 feet (3,000 m). | It hunts in almost total darkness for prey such as jumbo flying squid and giant squid. | Giant squid

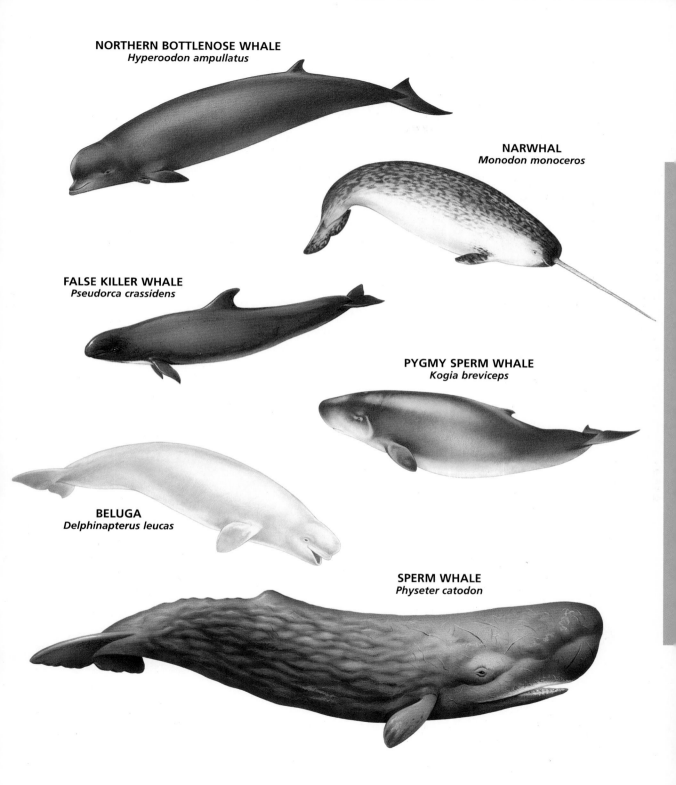

NORTHERN BOTTLENOSE WHALE
Hyperoodon ampullatus

NARWHAL
Monodon monoceros

FALSE KILLER WHALE
Pseudorca crassidens

PYGMY SPERM WHALE
Kogia breviceps

BELUGA
Delphinapterus leucas

SPERM WHALE
Physeter catodon

Dolphins

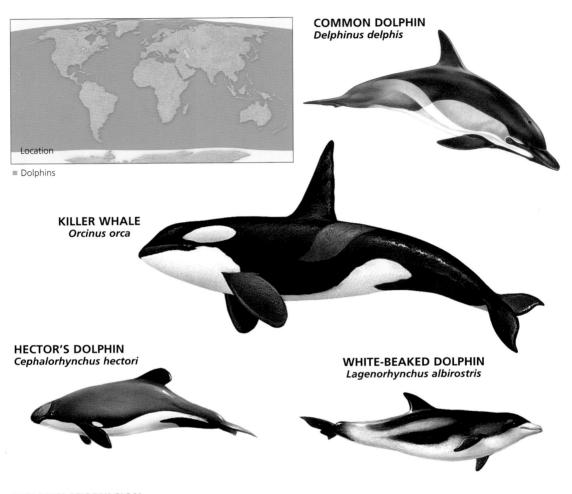

Location

■ Dolphins

COMMON DOLPHIN
Delphinus delphis

KILLER WHALE
Orcinus orca

HECTOR'S DOLPHIN
Cephalorhynchus hectori

WHITE-BEAKED DOLPHIN
Lagenorhynchus albirostris

DOLPHIN PROPULSION

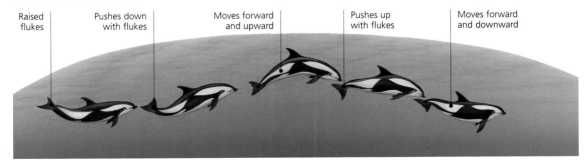

Raised flukes

Pushes down with flukes

Moves forward and upward

Pushes up with flukes

Moves forward and downward

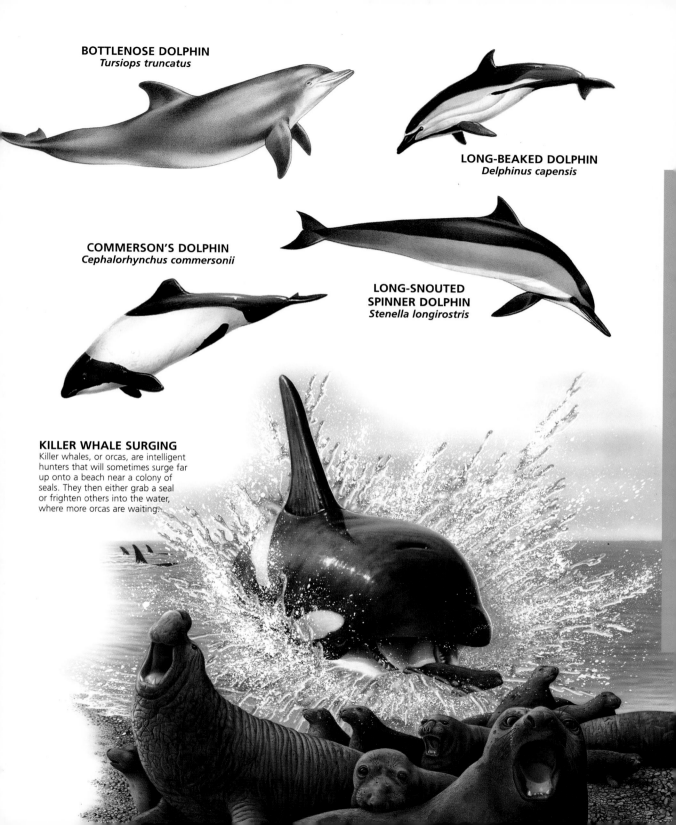

BOTTLENOSE DOLPHIN
Tursiops truncatus

LONG-BEAKED DOLPHIN
Delphinus capensis

COMMERSON'S DOLPHIN
Cephalorhynchus commersonii

**LONG-SNOUTED
SPINNER DOLPHIN**
Stenella longirostris

KILLER WHALE SURGING
Killer whales, or orcas, are intelligent
hunters that will sometimes surge far
up onto a beach near a colony of
seals. They then either grab a seal
or frighten others into the water,
where more orcas are waiting.

River Dolphins and Porpoises

Location

- River dolphins
- Porpoises
- River dolphins and porpoises

RIVER DOLPHIN ANATOMY

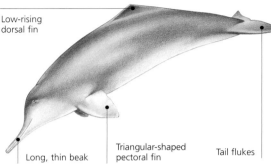

Low-rising dorsal fin

Long, thin beak

Triangular-shaped pectoral fin

Tail flukes

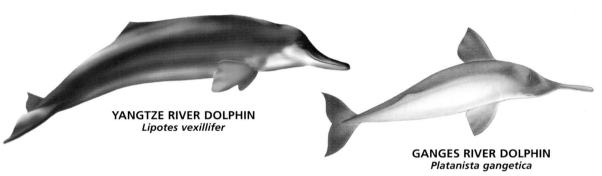

YANGTZE RIVER DOLPHIN
Lipotes vexillifer

GANGES RIVER DOLPHIN
Platanista gangetica

INDUS RIVER DOLPHIN
Platanista minor

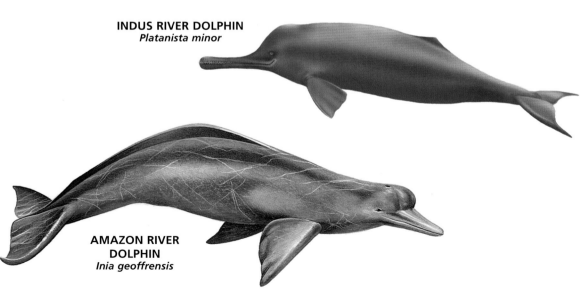

**AMAZON RIVER
DOLPHIN**
Inia geoffrensis

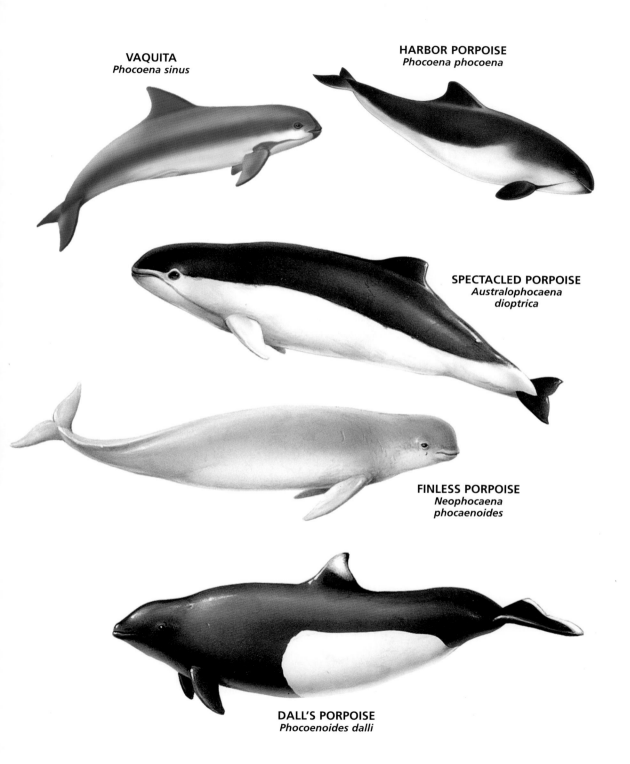

VAQUITA
Phocoena sinus

HARBOR PORPOISE
Phocoena phocoena

SPECTACLED PORPOISE
Australophocaena dioptrica

FINLESS PORPOISE
Neophocaena phocaenoides

DALL'S PORPOISE
Phocoenoides dalli

Hoofed Mammals

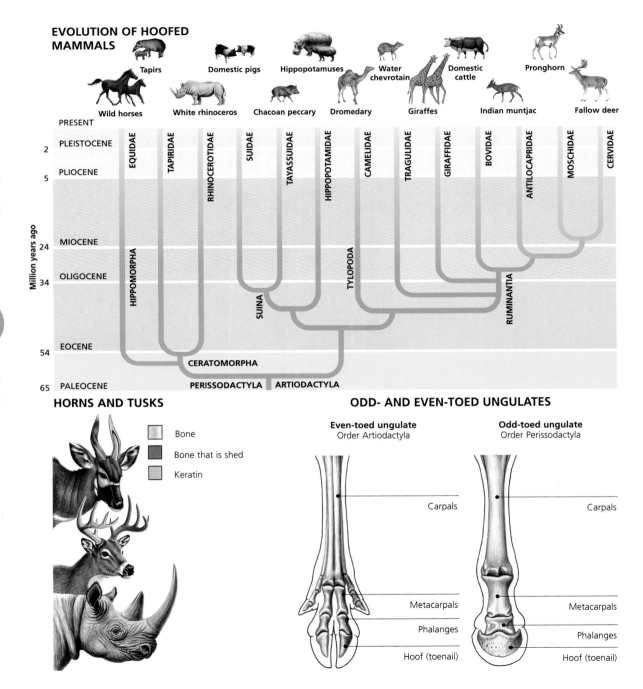

EVOLUTION OF HOOFED MAMMALS

Tapirs
Domestic pigs
Hippopotamuses
Water chevrotain
Domestic cattle
Pronghorn

Wild horses
White rhinoceros
Chacoan peccary
Dromedary
Giraffes
Indian muntjac
Fallow deer

Million years ago

PRESENT
PLEISTOCENE
2
PLIOCENE
5
MIOCENE
24
OLIGOCENE
34
EOCENE
54
PALEOCENE
65

EQUIDAE
TAPIRIDAE
RHINOCEROTIDAE
SUIDAE
TAYASSUIDAE
HIPPOPOTAMIDAE
CAMELIDAE
TRAGULIDAE
GIRAFFIDAE
BOVIDAE
ANTILOCAPRIDAE
MOSCHIDAE
CERVIDAE

HIPPOMORPHA
SUINA
TYLOPODA
RUMINANTIA

CERATOMORPHA
PERISSODACTYLA
ARTIODACTYLA

HORNS AND TUSKS

Bone

Bone that is shed

Keratin

ODD- AND EVEN-TOED UNGULATES

Even-toed ungulate
Order Artiodactyla

Carpals

Metacarpals

Phalanges

Hoof (toenail)

Odd-toed ungulate
Order Perissodactyla

Carpals

Metacarpals

Phalanges

Hoof (toenail)

Hoofed Mammals

MAMMALS

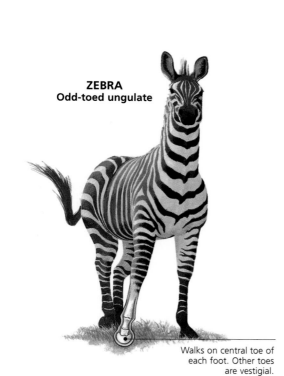

ZEBRA
Odd-toed ungulate

Walks on central toe of each foot. Other toes are vestigial.

DROMEDARY
Even-toed ungulate

Walks on third and fourth toes of each foot. No other toes remain.

WHITE RHINOCEROS
Odd-toed ungulate

Walks on middle three toes of front foot. No other toes remain.

REINDEER
Even-toed ungulate

Walks on four toes of each foot. No other toes remain.

179

Hippopotamuses, Pigs and Peccaries

HIPPOPOTAMUS HEAD

Location

- Wild pigs
- Peccaries
- Hippopotamuses and wild pigs

Fleshy lips used to pull out grass

Tusklike lower canine used for territorial fighting

Lower incisor

Eyes, ears and nostrils located on top of head

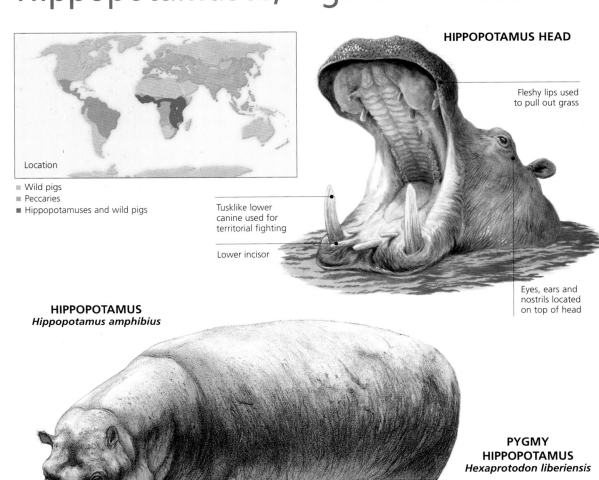

HIPPOPOTAMUS
Hippopotamus amphibius

PYGMY HIPPOPOTAMUS
Hexaprotodon liberiensis

CHACOAN PECCARY
Catagonus wagneri

DOMESTIC PIG
Sus scrofa

BABIRUSA
Babyrousa babyrussa

RED RIVER HOG
Potamochoerus porcus

BEARDED PIG
Sus barbatus

WARTHOG
Phacochoerus africanus

Old World Camels

Location

■ Bactrian camels (wild dromedary camels extinct)

CAMEL HUMP

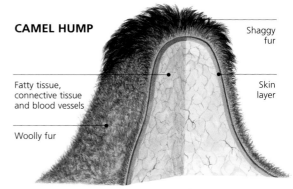

Shaggy fur

Fatty tissue, connective tissue and blood vessels

Skin layer

Woolly fur

CAMEL ANATOMY

Horny padding on hindlegs protects camels from heat.

Nostrils can be closed in sandstorms.

Camels walk two legs at a time, right following left, which causes their body to rock and sway.

Two rows of eyelashes keep out sand.

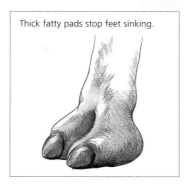

Thick fatty pads stop feet sinking.

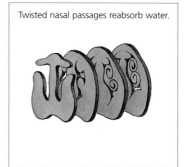

Twisted nasal passages reabsorb water.

DROMEDARY
Camelus dromedarius

BACTRIAN CAMEL
Camelus bactrianus

Arabian camel-racing

Algerian nomad with camel

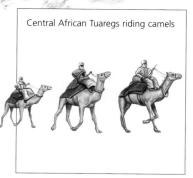

Central African Tuaregs riding camels

New World Camelids

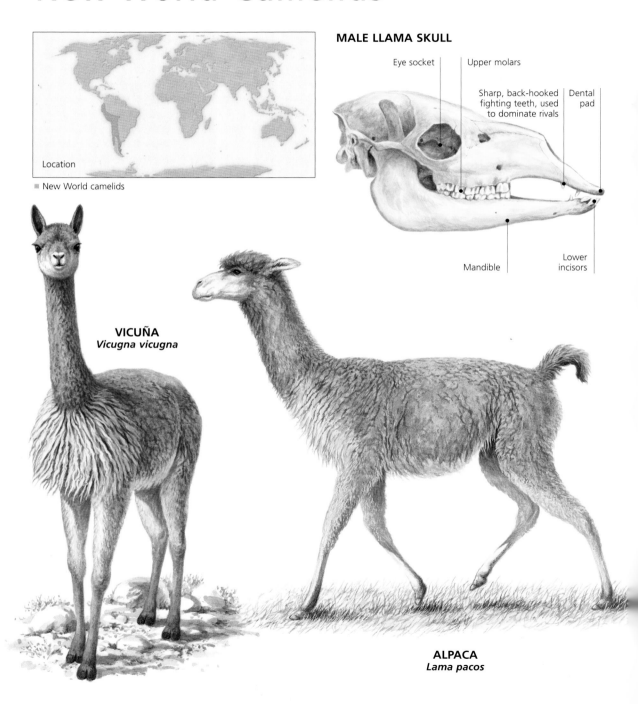

Location

■ New World camelids

MALE LLAMA SKULL

Eye socket

Upper molars

Sharp, back-hooked
fighting teeth, used
to dominate rivals

Dental
pad

Mandible

Lower
incisors

VICUÑA
Vicugna vicugna

ALPACA
Lama pacos

LLAMA
Lama glama

**GUANACO
PREPARING TO SPIT**
Lama guanicoe

Domestic alpaca ready for shearing

Guancos alert
for predators

Domestic llamas bred as pack animals

Giraffe, Okapi and Pronghorn

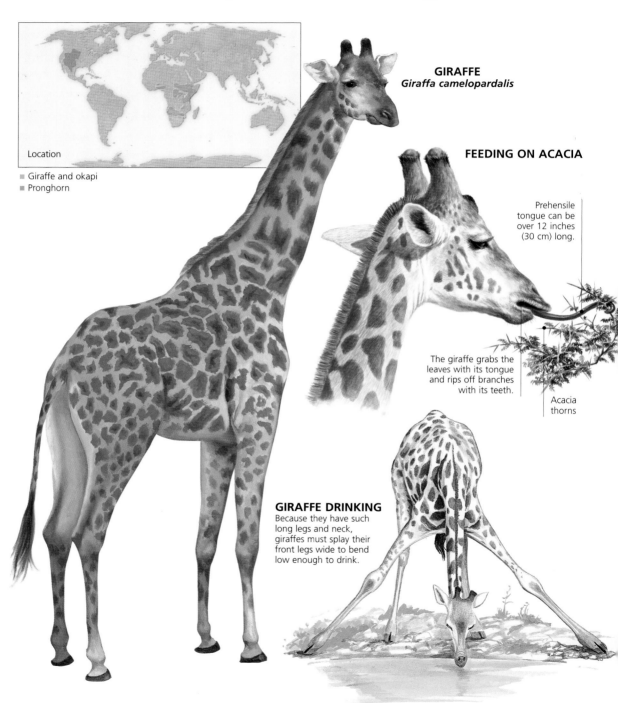

Location

■ Giraffe and okapi
■ Pronghorn

GIRAFFE
Giraffa camelopardalis

FEEDING ON ACACIA

Prehensile tongue can be over 12 inches (30 cm) long.

The giraffe grabs the leaves with its tongue and rips off branches with its teeth.

Acacia thorns

GIRAFFE DRINKING
Because they have such long legs and neck, giraffes must splay their front legs wide to bend low enough to drink.

PRONGHORN TAKING FLIGHT
The fastest animal in the New World, the pronghorn flashes its white underside in warning as it bounds away from danger.

PRONGHORN
Antilocapra americana

OKAPI
Okapia johnstoni

Deer

SPREADING MOOSE ANTLERS

One year
Horns are simple spikes.

Four years
Rack expands in size
and number of points.

Eight years
Mature rack demonstrates moose's
suitability as a potential mate.

MALE CARIBOU FIGHTING

Male caribou fight by
locking antlers during
the breeding season.

Fighting males rarely
become entangled by
their racks of antlers.

After pushing to see which is
stonger, the victor wins the right
to mate with the herd's females.

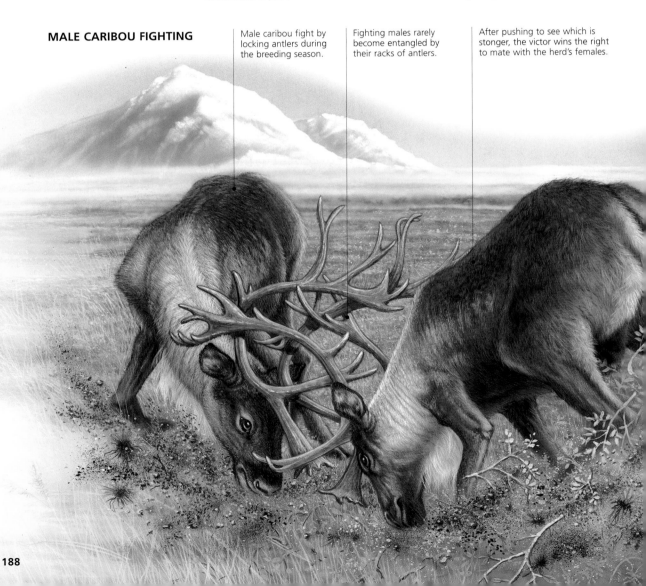

188

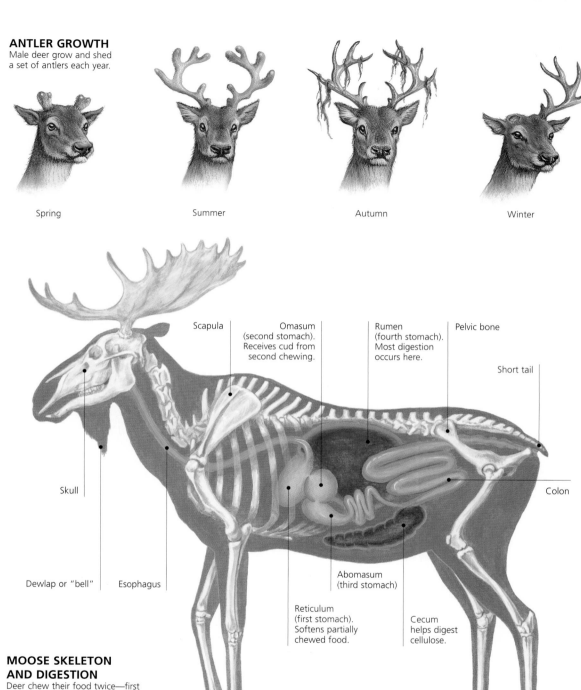

ANTLER GROWTH
Male deer grow and shed a set of antlers each year.

Spring

Summer

Autumn

Winter

Scapula

Omasum (second stomach). Receives cud from second chewing.

Rumen (fourth stomach). Most digestion occurs here.

Pelvic bone

Short tail

Skull

Colon

Dewlap or "bell"

Esophagus

Abomasum (third stomach)

Reticulum (first stomach). Softens partially chewed food.

Cecum helps digest cellulose.

MOOSE SKELETON AND DIGESTION
Deer chew their food twice—first taking large bites of food, then chewing the partially digested 'cud' in safety. This process allows a large amount of nutrient-poor food to be eaten quickly but digested slowly.

Deer

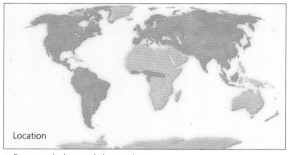

Location

■ Deer, musk deer and chevrotains

PUDU
Pudu sp.

WATER CHEVROTAIN
Hyemoschus aquaticus

BARASINGHA
Cervus duvaucelii

CARIBOU
Rangifer tarandus

INDIAN MUNTJAC
Muntiacus muntjak

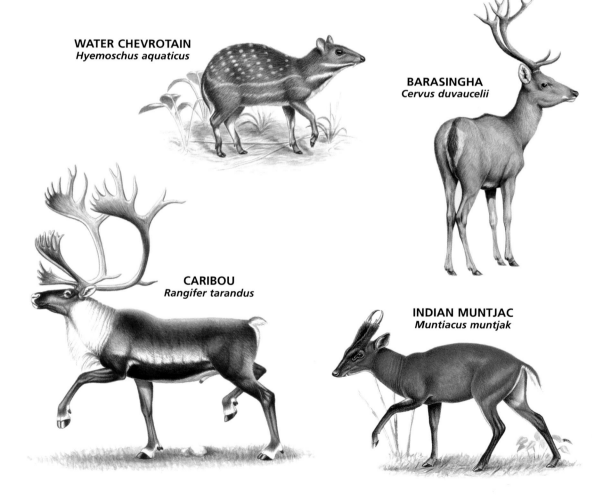

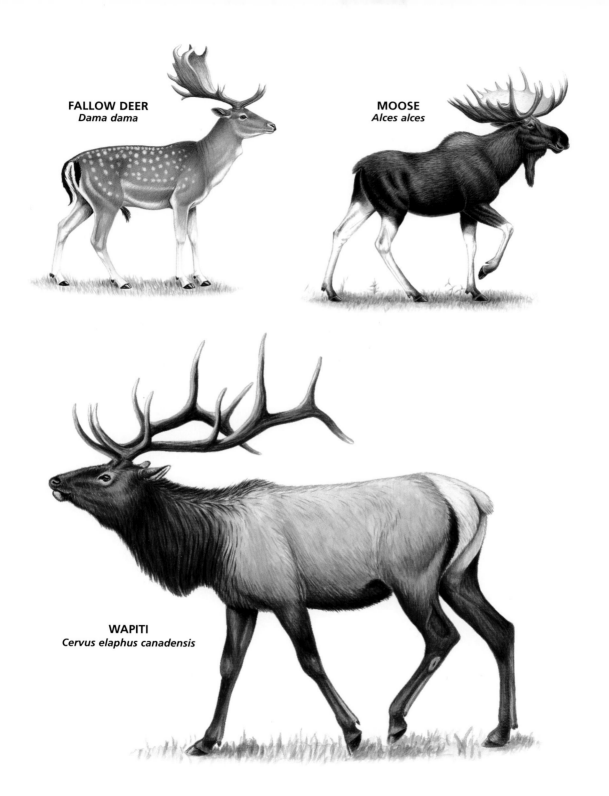

FALLOW DEER
Dama dama

MOOSE
Alces alces

WAPITI
Cervus elaphus canadensis

191

Bovids

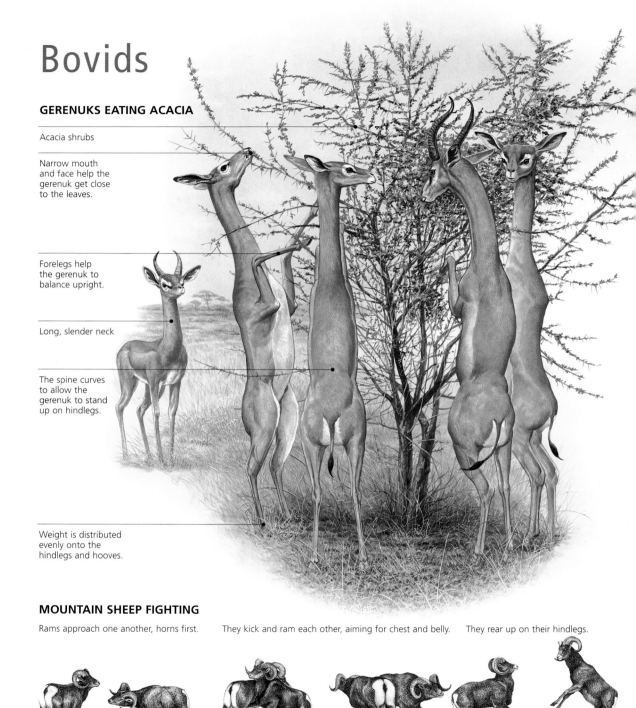

GERENUKS EATING ACACIA

Acacia shrubs

Narrow mouth and face help the gerenuk get close to the leaves.

Forelegs help the gerenuk to balance upright.

Long, slender neck

The spine curves to allow the gerenuk to stand up on hindlegs.

Weight is distributed evenly onto the hindlegs and hooves.

MOUNTAIN SHEEP FIGHTING

Rams approach one another, horns first.　　They kick and ram each other, aiming for chest and belly.　　They rear up on their hindlegs.

TYPES OF BOVIDS

Barbary sheep
Goat

Gaur
True cattle

Greater kudu
Antelope

Then drop with great force into a head-on clash.

They then present to each other.

The rams break from fighting to feed.

Antelopes

Location

■ Antelopes, gazelles and duikers

SABLE ANTELOPE
Hippotragus niger

DUIKER
Cephalophus sp.

KLIPSPRINGER
Oreotragus oreotragus

BONGO
Tragelaphus eurycerus

BLUE WILDEBEEST
Connochaetus taurinus

THOMSON'S GAZELLE
Gazella thomsonii

SAIGA ANTELOPE
Saiga tatarica

GEMSBOK
Oryx gazella

GERENUK
Litocranius walleri

FOUR-HORNED
ANTELOPE
Tetracerus quadricornis

BONTEBOK
Damaliscus pygargus

BLACKBUCK
Antilope cervicapra

195

Wildebeest Migration

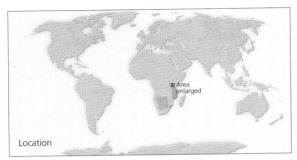

Location

■ Wildebeest

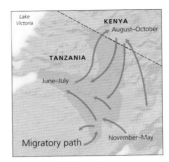

■ Serengeti National Park
■ Migration through Serengeti–Mara ecosystem

ACROSS THE SERENGETI

Each year, thousands of wildebeest make their way across the Serengeti–Mara ecosystem in Tanzania and Kenya, a journey that can be 1,900 miles (3,000 km) long. In the dry season, the wildebeest move from acacia woodlands onto the Serengeti Plain in search of water and grass. They return in the wet season.

Calves are born at the beginning of the rainy season, and stay with their mothers until eight months of age, when they leave to form peer-group herds.

Wildebeest never stay in one spot long enough to damage the environment, and as the columns of animals move forward, the soil is fertilized by their droppings.

Zebras migrate across the Serengeti at the same time, eating grasses cropped short by the wildebeest, and using the tens of thousands of animals as protection from predators.

River crossings, such as the Mara River, can be deadly, with crocodiles and other predators lying in wait for an animal to stumble or drift away from the mass.

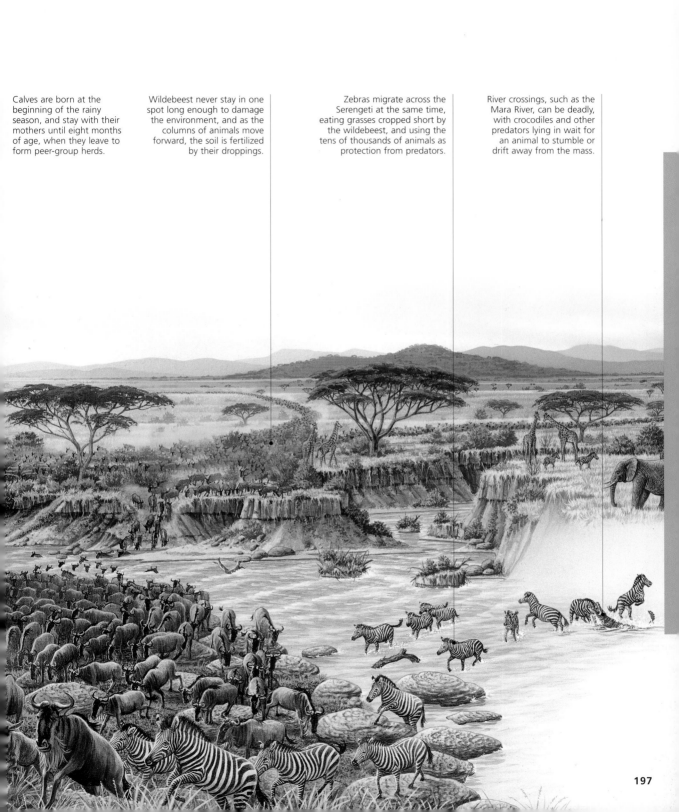

True Cattle, Goats and Sheep

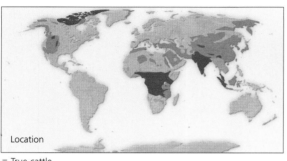

Location

■ True cattle
■ Goats and sheep
■ True cattle, goats and sheep

YAK
Bos grunniens

MOUNTAIN ANOA
Bubalus quarlesi

AMERICAN BISON
Bison bison

EUROPEAN BISON
Bison bonasus

AFRICAN BUFFALO
Syncerus caffer

DOMESTIC BREEDS OF CATTLE

Hereford cattle

Friesian cattle

Galloway cattle

Brahmin cattle

Ankole cattle

Texas long-horned cattle

MUSKOX
Ovibos moschatus

MOUFLON
Ovis orientalis musimon

DALL'S SHEEP
Ovis dalli

BIGHORN SHEEP
Ovis canadensis

HIMALAYAN TAHR
Hemitragus jemlahicus

WILD GOAT
Capra aegagrus

MOUNTAIN GOAT
Oreamnos americanus

DOMESTIC BREEDS OF SHEEP AND GOATS

Leicester sheep

Cashmere goat

Lincoln sheep

Black-faced sheep

Merino sheep

Domestic sheep

Romney sheep

199

Rhinoceroses

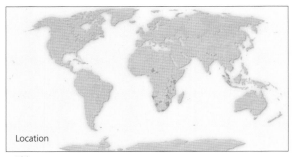

Location

■ Rhinoceroses

RHINO DEFENSE
Groups of rhinoceroses form a circle facing outward to protect their calves.

BLACK RHINO CHARGING

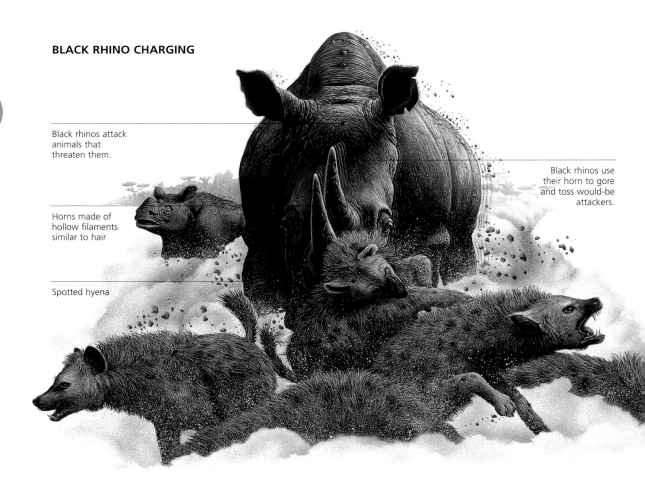

Black rhinos attack animals that threaten them.

Horns made of hollow filaments similar to hair

Spotted hyena

Black rhinos use their horn to gore and toss would-be attackers.

JAVAN RHINOCEROS
Rhinoceros sondaicus

INDIAN RHINOCEROS
Rhinoceros unicornis

SUMATRAN RHINOCEROS
Dicerorhinus sumatrensis

WHITE RHINOCEROS
Ceratotherium simum

BLACK RHINOCEROS
Diceros bicornis

RHINO COMPARISON

Indian rhinoceros
Asian one-horned

Black rhinoceros
African two-horned

Horses, Asses and Zebras

HORSE GALLOPING
When horses gallop, each hoof strikes the ground separately.

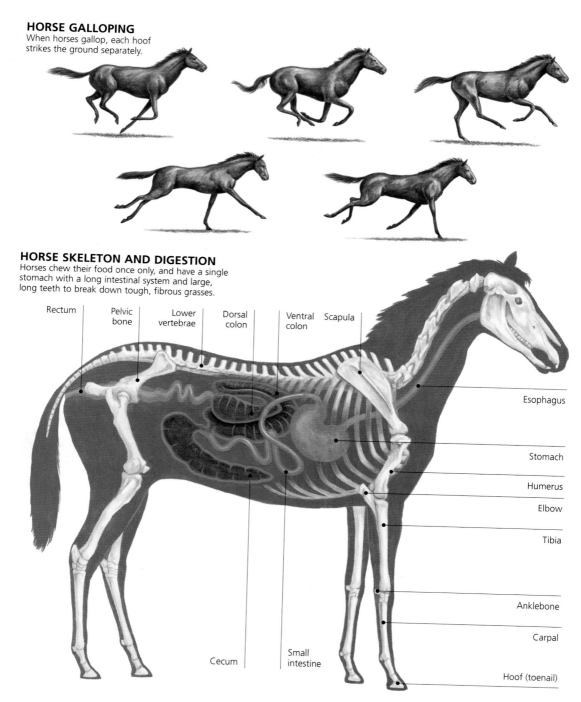

HORSE SKELETON AND DIGESTION
Horses chew their food once only, and have a single stomach with a long intestinal system and large, long teeth to break down tough, fibrous grasses.

Rectum

Pelvic bone

Lower vertebrae

Dorsal colon

Ventral colon

Scapula

Esophagus

Stomach

Humerus

Elbow

Tibia

Anklebone

Carpal

Hoof (toenail)

Cecum

Small intestine

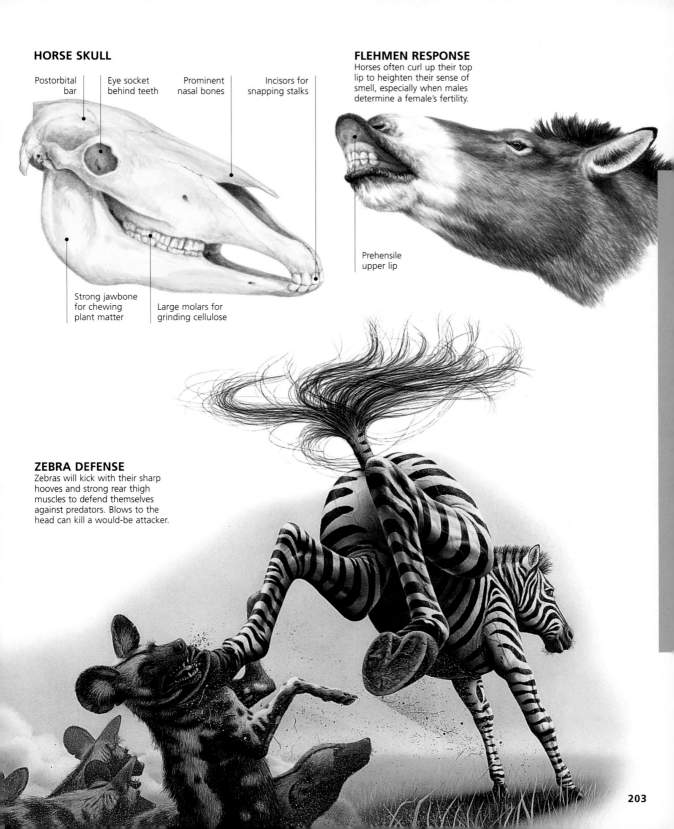

HORSE SKULL

Postorbital bar

Eye socket behind teeth

Prominent nasal bones

Incisors for snapping stalks

Strong jawbone for chewing plant matter

Large molars for grinding cellulose

FLEHMEN RESPONSE

Horses often curl up their top lip to heighten their sense of smell, especially when males determine a female's fertility.

Prehensile upper lip

ZEBRA DEFENSE

Zebras will kick with their sharp hooves and strong rear thigh muscles to defend themselves against predators. Blows to the head can kill a would-be attacker.

Horses, Asses and Zebras

Location

■ Wild horses, asses and zebras

HORSES AND PEOPLE

Training Lipizzaner horses

Canadian mountie

Lokai horse and horseman

Cowboy lassooing cattle

Playing polo

Racing horses

Austrian sled racing

KIANG
Equus kiang

ONAGER
Equus onager

ASS
Equus asinus

HORSE
Equus caballus

TARPAN
Equus caballus gmelini

ZEBRA
Equus sp.

DOMESTIC HORSES

Feral horses

Plains pony

Arabian horse

Aardvark, Hyraxes and Tapirs

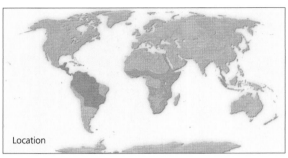

Location

■ Aardvark and hyraxes
■ Tapirs

YELLOW-SPOTTED HYRAX
Heterohyrax brucei

ROCK HYRAXES
Procavia capensis

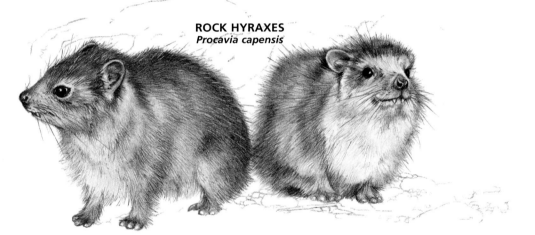

AARDVARK
Orycteropus afer

AARDVARK ANATOMY

Tops of ears can fold back to protect them when digging.

Aardvarks sniff out ants and termites to eat.

MOUNTAIN TAPIR
Tapirus pinchaque

BAIRD'S TAPIR
Tapirus bairdii

SOUTH AMERICAN TAPIR
WITH YOUNG
Tapirus terrestris

MALAYAN TAPIR
Tapirus indicus

Elephants

ELEPHANT SKIN

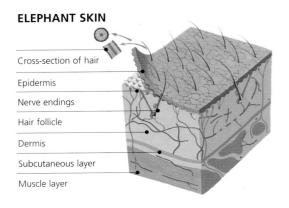

Cross-section of hair

Epidermis

Nerve endings

Hair follicle

Dermis

Subcutaneous layer

Muscle layer

ELEPHANT FOOT

Elephants walk on toes of feet.

Fibrous pads support the toes.

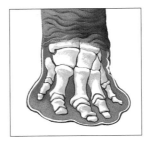

ASIAN ELEPHANT

Domed forehead

Ears shorter than neck

Convex or level back

One finger on tip of trunk

Skin relatively smooth

AFRICAN ELEPHANT

No dome on forehead

Ears longer than neck

Concave back

Two fingers on tip of trunk

Skin wrinkled and baggy

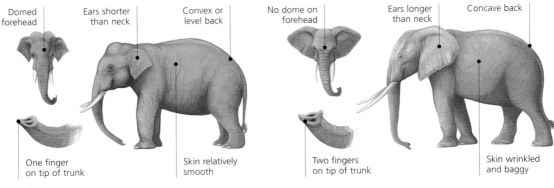

ELEPHANT GROWTH

Less than one year

One year

Three years

Six years

10 years

15 years

40 years

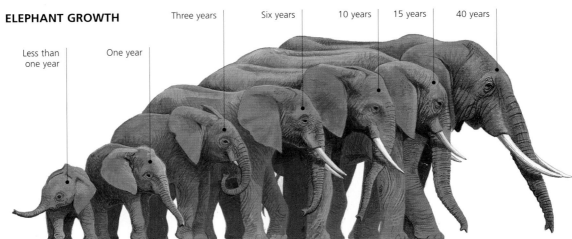

AFRICAN ELEPHANT

AFRICAN ELEPHANT SKELETON

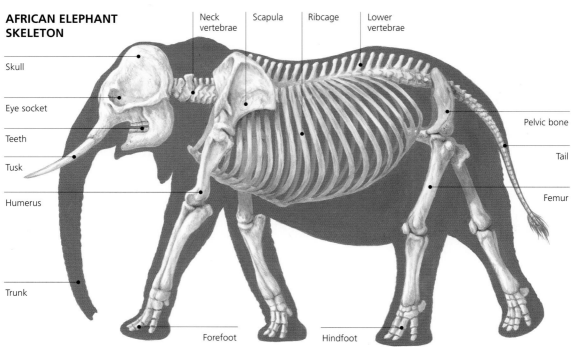

Neck vertebrae

Scapula

Ribcage

Lower vertebrae

Skull

Eye socket

Teeth

Tusk

Humerus

Trunk

Pelvic bone

Tail

Femur

Forefoot

Hindfoot

Elephants

Location

■ Elephants

ELEPHANT TUSKS

Tusks have c
containing n
endings, whi
pressure-sens

AFRICAN BUSH ELEPHANTS
Loxodonta africana

SUMATRAN ASIAN ELEPHANTS
Elephas maximus sumatranus

SRI LANKAN ASIAN ELEPHANTS
Elephas maximus maximus

MAINLAND ASIAN ELEPHANTS
Elephas maximus indicus

FOREST AFRICAN ELEPHANTS
Loxodonta cyclotis

Dugong and Manatees

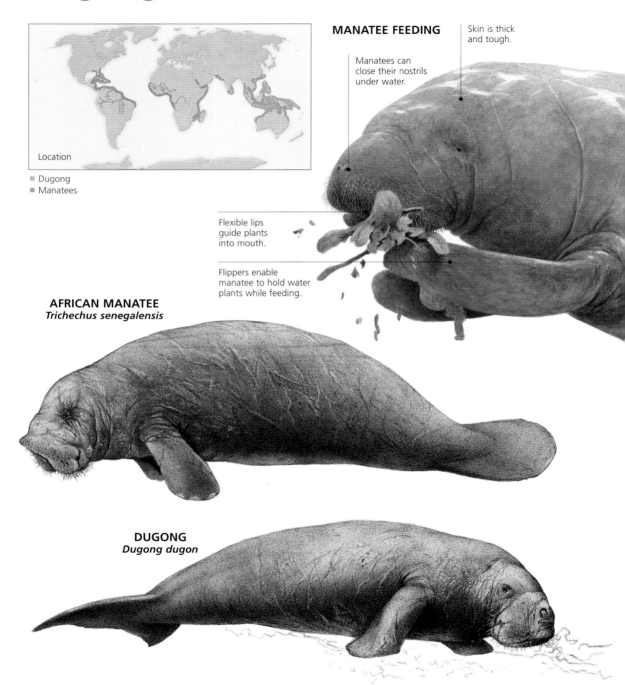

Location
- Dugong
- Manatees

MANATEE FEEDING

Skin is thick
and tough.

Manatees can
close their nostrils
under water.

Flexible lips
guide plants
into mouth.

Flippers enable
manatee to hold water
plants while feeding.

AFRICAN MANATEE
Trichechus senegalensis

DUGONG
Dugong dugon

WEST INDIAN MANATEE
Trichechus manatus

AMAZONIAN MANATEE
Trichechus inunguis

Elephant Shrews

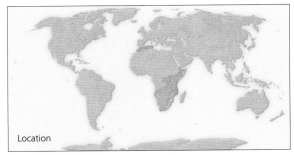

Location

■ Elephant shrews

ELEPHANT SHREW HEAD

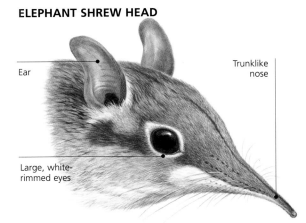

Ear

Large, white-rimmed eyes

Trunklike nose

ROCK ELEPHANT SHREW HABITAT
Elephantulus myurus

Rock elephant shrews nest in the crevices of rock overhangs to protect themselves from predators and the elements.

They have large eyes and a long, sensitive and mobile snout, which is used to find insects.

Rock elephant shrews are generally soft gray-brown in color, which acts as camouflage from predators.

SHORT-EARED ELEPHANT SHREW
Macroscelides proboscideus

GOLDEN-RUMPED ELEPHANT SHREW
Rhynchocyon chrysopygus

RUFOUS ELEPHANT SHREW
Elephantulus rufescens

**NORTH AFRICAN
ELEPHANT SHREW**
Elephantulus rozeti

FOUR-TOED ELEPHANT SHREW
Petrodromus tetradactylus

215

Rodents

EVOLUTION OF RODENTS

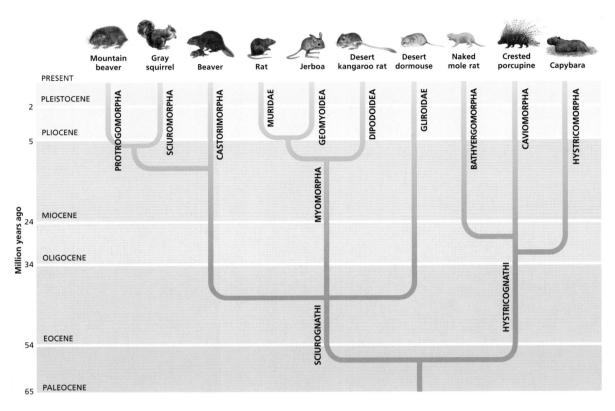

Mountain beaver	Gray squirrel	Beaver	Rat	Jerboa	Desert kangaroo rat	Desert dormouse	Naked mole rat	Crested porcupine	Capybara

PRESENT
PLEISTOCENE
2
PLIOCENE
5
MIOCENE
24
OLIGOCENE
34
EOCENE
54
PALEOCENE
65

Million years ago

PROTROGOMORPHA
SCIUROMORPHA
CASTORIMORPHA
MURIDAE
GEOMYOIDEA
MYOMORPHA
DIPODOIDEA
GLIROIDEA
BATHYERGOMORPHA
HYSTRICOGNATHI
CAVIOMORPHA
HYSTRICOMORPHA
SCIUROGNATHI

Rodents

MAMMALS

RODENT SKULL

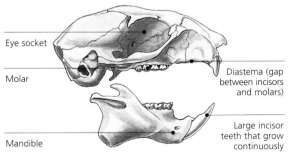

Eye socket

Molar

Mandible

Diastema (gap between incisors and molars)

Large incisor teeth that grow continuously

RODENT HEAD

Ear

Cheeks able to store food while gathering

Mandible

Excellent vision

Whiskers

Lower incisor

RODENT JAW MUSCLES

■ Lateral jaw muscles ■ Deep jaw muscles

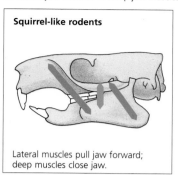

Squirrel-like rodents

Lateral muscles pull jaw forward; deep muscles close jaw.

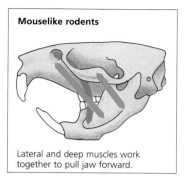

Mouselike rodents

Lateral and deep muscles work together to pull jaw forward.

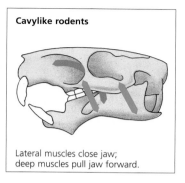

Cavylike rodents

Lateral muscles close jaw; deep muscles pull jaw forward.

EURASIAN RED SQUIRREL
Squirrel-like rodent

PREHENSILE-TAILED PORCUPINE
Cavylike rodent

**MEXICAN
SPINY POCKET MOUSE**
Mouselike rodent

LEAST CHIPMUNK
Squirrel-like rodent

MOUNTAIN BEAVER
Squirrel-like rodent

Squirrel-like Rodents

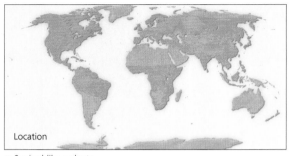

Location

■ Squirrel-like rodents

PRAIRIE DOGS
Cynomys **sp.**

PRAIRIE DOG COMMUNITY

Prairie dogs greet each other with kissing ceremonies.

Guards whistle an alarm when in danger.

Ferrets can crawl through the tunnels in search of prey.

Females build grass nests safe underground.

Members of different coteries greet each other by sniffing.

Prairie dogs live in social units called coteries. A number of interconnected coteries form a town.

Snakes and burrowing owls often take over unused burrows.

EASTERN FOX SQUIRREL
Sciurus niger

DOUGLAS' SQUIRREL
Tamiasciurus douglasii

MARMOT
Marmota sp.

WOODCHUCK
Marmota monax

GRAY SQUIRREL
Sciurus sp.

ABERT'S SQUIRREL
Sciurus aberti

EASTERN CHIPMUNK
Tamias striatus

Squirrel-like Rodents

BLACK GIANT SQUIRREL
Ratufa bicolor

DELMARVA FOX SQUIRREL
Sciurus niger cenerus

AMERICAN RED SQUIRREL
Tamiasciurus hudsonicus

LEAST CHIPMUNK
Tamias minimus

HOARY MARMOT
Marmota caligata

COLUMBIAN GROUND SQUIRREL
Spermophilus columbianus

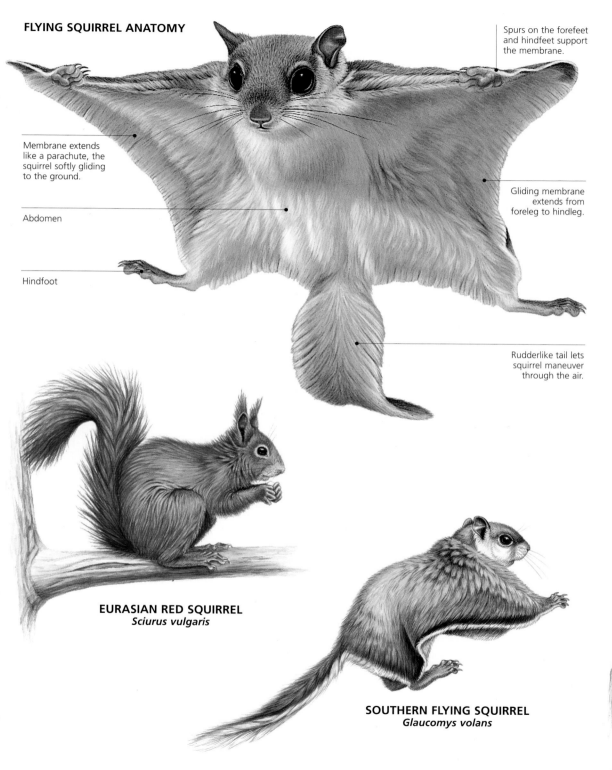

FLYING SQUIRREL ANATOMY

Spurs on the forefeet and hindfeet support the membrane.

Membrane extends like a parachute, the squirrel softly gliding to the ground.

Gliding membrane extends from foreleg to hindleg.

Abdomen

Hindfoot

Rudderlike tail lets squirrel maneuver through the air.

EURASIAN RED SQUIRREL
Sciurus vulgaris

SOUTHERN FLYING SQUIRREL
Glaucomys volans

Beavers

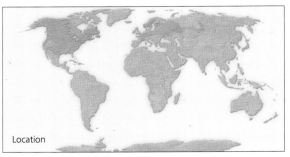

Location

■ Beavers and mountain beavers

AMERICAN BEAVER AND KIT
Castor canadensis

BEAVER DAM AND LODGE

Beavers eat bark and use their strong incisor teeth to chew through tree trunks.

After felling trees and saplings for cross-braces, beavers carry smaller logs in their mouths.

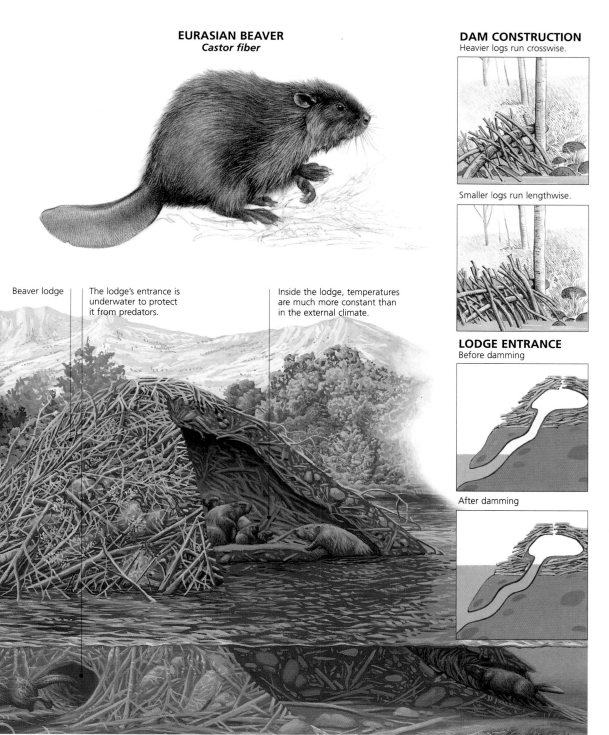

EURASIAN BEAVER
Castor fiber

Beaver lodge

The lodge's entrance is underwater to protect it from predators.

Inside the lodge, temperatures are much more constant than in the external climate.

DAM CONSTRUCTION
Heavier logs run crosswise.

Smaller logs run lengthwise.

LODGE ENTRANCE
Before damming

After damming

Mouselike Rodents

Location

■ Mouselike rodents

MUSKRAT
Ondatra zibethicus

MASKED MOUSE-TAILED DORMOUSE
Myomimus personatus

DESERT DORMOUSE
Selevinia betpakdalaensis

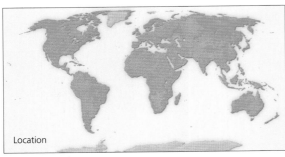

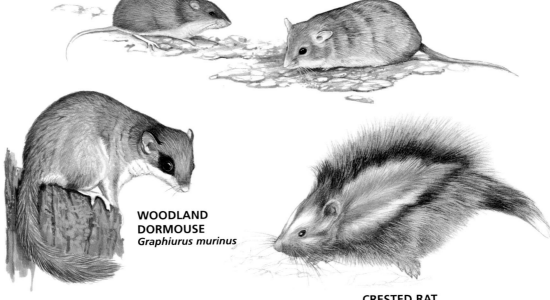

**WOODLAND
DORMOUSE**
Graphiurus murinus

CRESTED RAT
Lophiomys imhausi

LEMMING
Lemmus sp.

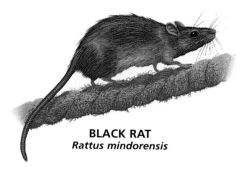

BLACK RAT
Rattus mindorensis

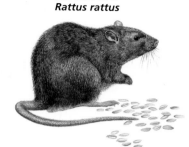

HOUSE RAT
Rattus rattus

LONG-TAILED POCKET MOUSE
Chaetodipus formosus

GERBIL
Gerbillus sp.

JERBOA
Jaculus sp.

BANK VOLE
Clethrionomys glareolus

GRASSHOPPER MOUSE
Onychomys sp.

DESERT KANGAROO RAT
Dipodomys deserti

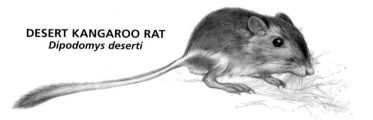

Cavylike Rodents

BRAZILIAN PORCUPINE
Coendou prehensilis

Location

■ Cavylike rodents

CAPYBARA
Hydrochaeris hydrochaeris

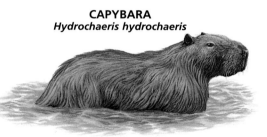

BRAZILIAN AGOUTI
Dasyprocta leporina

CRESTED PORCUPINE
Hystrix cristata

MALAYAN PORCUPINE
Hystrix brachyura

CORURO
Spalacopus cyanus

GUINEA PIG
Cavia aperea

CHILEAN ROCK RAT
Aconaemys fuscus

PLAINS VISCACHA
Lagostomus maximus

CHINCHILLA
Chinchilla lanigera

PATAGONIAN MARA
Dolichotis patagonum

SPINY RAT
Family *Echimyidae*

Rabbits, Hares and Pikas

BLACK-LIPPED PIKA
Ochotona curzoniae
Pikas are short-legged and short-eared
relatives of rabbits and hares. Mainly
herbivorous, they live on the steppes
and plains of North America and
continental Asia.

RABBIT ANATOMY

Long ears to detect
approach of predators
and to keep cool

Curved skull

Eyes at side of head to
locate predators from
above and behind

Continuously
growing incisors

RABBIT WARREN

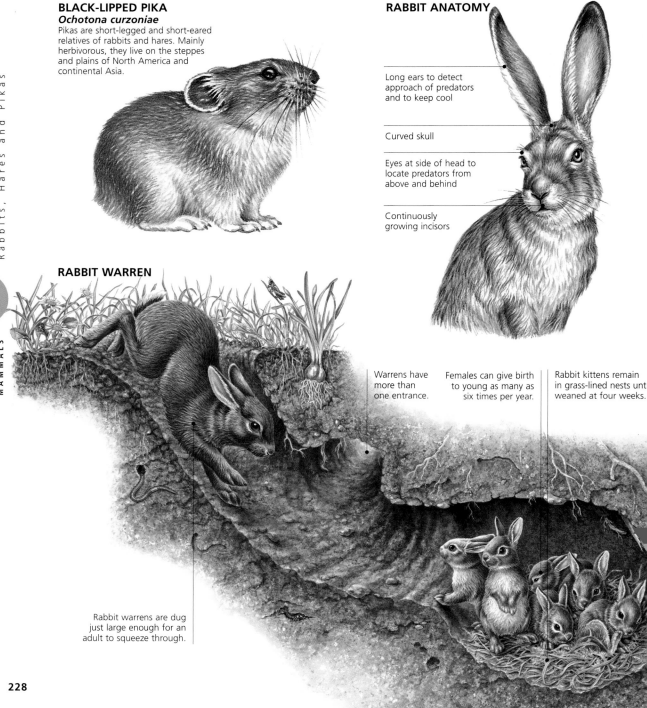

Warrens have
more than
one entrance.

Females can give birth
to young as many as
six times per year.

Rabbit kittens remain
in grass-lined nests unt
weaned at four weeks.

Rabbit warrens are dug
just large enough for an
adult to squeeze through.

SNOWSHOE HARE ADAPTATIONS

Snowshoe hares eat treebark in winter to survive.

White winter coat

Huge feet covered in long, stiff hairs distribute weight.

Comparatively short ears minimize heat loss.

Brown summer coat

Summer diet of green vegetation and berries

RABBIT ADAPTATIONS

Arctic hare
Small extremities and thick pelt reduce the body's surface area to conserve heat.

Antelope jackrabbit
Long extremities and small body core enlarge its surface area to dispel heat.

Black-tailed jackrabbit
Hundreds of blood vessels in the ears radiate heat away from the body.

Rabbits, Hares and Pikas

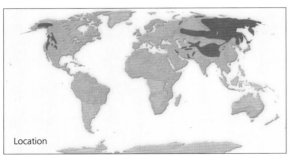

Location

- Rabbits and hares
- Rabbits, hares and pikas

SUMATRAN RABBIT
Nesolagus netscheri

RYUKYU RABBIT
Pentalagus furnessi

VOLCANO RABBIT
Romerolagus diazi

BRUSH RABBIT
Sylvilagus bachmani

RIVENNE RABBIT
Bunolagus monticularis

PYGMY RABBIT
Brachylagus idahoensis

NORTHERN PIKA
Ochotona alpina

HISPID HARE
Caprolagus hispidus

CENTRAL AFRICAN HARE
Poelagus marjorita

ROYLE'S PIKA
Ochotona roylei

AMERICAN PIKA
Ochotona princeps

PIKA BURROW

Nostrils close against the cold weather.

Long, thick fur and no visible tail

Fur on soles of feet

Pikas live in rock crevices and under overhangs.

Piles of grass and flowers are stored to eat over winter.

Pikas live among the rocks, whistling to warn others of predators.

233

Classifying Birds

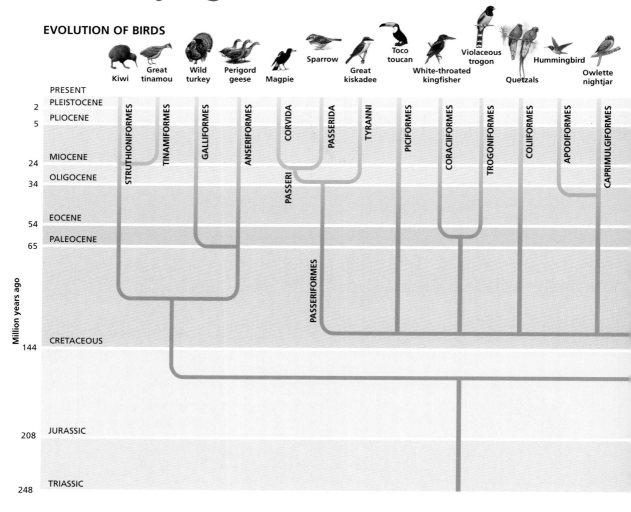

EVOLUTION OF BIRDS

Kiwi · Great tinamou · Wild turkey · Perigord geese · Magpie · Sparrow · Great kiskadee · Toco toucan · White-throated kingfisher · Violaceous trogon · Quetzals · Hummingbird · Owlette nightjar

STRUTHIONIFORMES · TINAMIFORMES · GALLIFORMES · ANSERIFORMES · PASSERI · CORVIDA · PASSERIDA · TYRANNI · PICIFORMES · CORACIIFORMES · TROGONIFORMES · COLIIFORMES · APODIFORMES · CAPRIMULGIFORMES

PASSERIFORMES

Million years ago

PRESENT	
PLEISTOCENE	2
PLIOCENE	5
MIOCENE	24
OLIGOCENE	34
EOCENE	54
PALEOCENE	65
CRETACEOUS	144
JURASSIC	208
TRIASSIC	248

Male (top) and female gambel's quails

Bonellis eagle

Painted stork

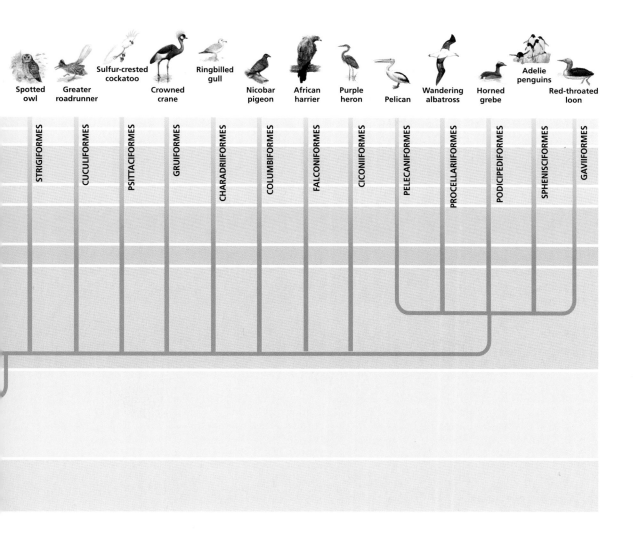

Spotted owl

Greater roadrunner

Sulfur-crested cockatoo

Crowned crane

Ringbilled gull

Nicobar pigeon

African harrier

Purple heron

Pelican

Wandering albatross

Horned grebe

Adelie penguins

Red-throated loon

STRIGIFORMES

CUCULIFORMES

PSITTACIFORMES

GRUIFORMES

CHARADRIIFORMES

COLUMBIFORMES

FALCONIFORMES

CICONIIFORMES

PELECANIFORMES

PROCELLARIIFORMES

PODICIPEDIFORMES

SPHENISCIFORMES

GAVIIFORMES

Oilbird

Tree swallow

Cedar waxwing

Bird Characteristics

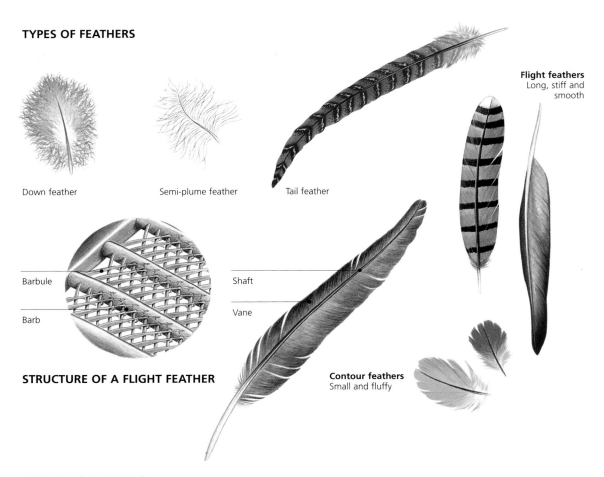

TYPES OF FEATHERS

Down feather

Semi-plume feather

Tail feather

Flight feathers
Long, stiff and smooth

Barbule

Barb

Shaft

Vane

STRUCTURE OF A FLIGHT FEATHER

Contour feathers
Small and fluffy

REPLACING FEATHERS

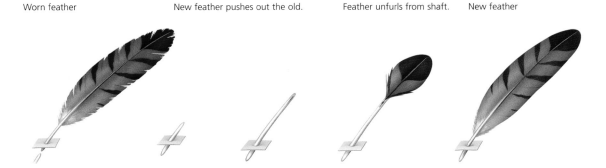

Worn feather

New feather pushes out the old.

Feather unfurls from shaft.

New feather

Introducing Birds

BIRDS

PLUMAGE FEATHERS

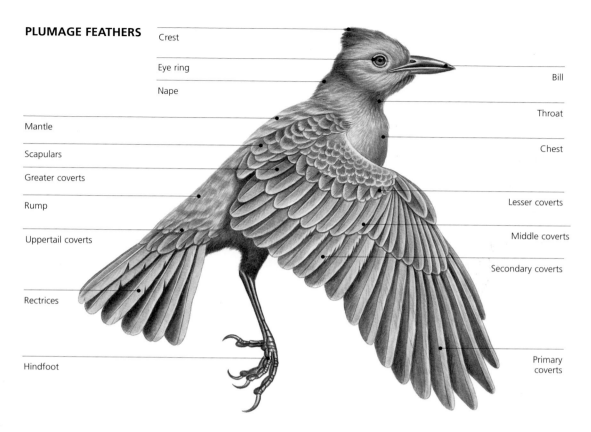

Crest

Eye ring

Nape

Mantle

Scapulars

Greater coverts

Rump

Uppertail coverts

Rectrices

Hindfoot

Bill

Throat

Chest

Lesser coverts

Middle coverts

Secondary coverts

Primary coverts

INSIDE AN EGG

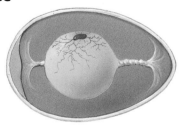

Newly laid egg
Small embryo
uses yolk and
albumen as food.

Embryonic chick
As embryo grows,
wastes are stored
in a special sac.

Developing chick
When close to
hatching, chick
nearly fills egg.

Ready to hatch
Unhatched chicks
use an egg tooth
to break free.

Evolution of Birds

HESPERORNIS
Ancestor to modern grebes
that lived 70 million years ago

ARCHAEOPTERYX
The earliest known bird, it
lived 145 million years ago.

Waterfowl ancestor
Ichthyornis

Earliest bird
Archaeopteryx

Dinosaur ancestor
Reptilian theropod

TERROR-BIRD
At 9 feet (2.7 m), the largest of all
known birds lived 1 million years ago.

Introducing Birds

BIRDS

240

EVOLUTION OF MODERN BIRDS

Modern waterfowl
Snow goose

DINORNIS MAXIMUS
Recently extinct flightless bird
that lived 200–300 years ago

TERATORNIS MERRIAMI
An ancestor of modern vultures,
it lived 2 million years ago.

Bird Anatomy

AVIAN ANATOMY

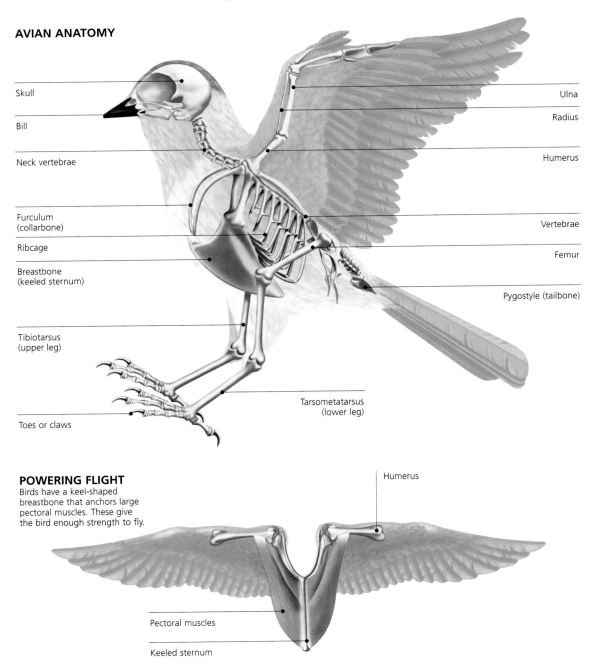

Skull

Bill

Neck vertebrae

Furculum
(collarbone)

Ribcage

Breastbone
(keeled sternum)

Tibiotarsus
(upper leg)

Toes or claws

Ulna

Radius

Humerus

Vertebrae

Femur

Pygostyle (tailbone)

Tarsometatarsus
(lower leg)

POWERING FLIGHT

Birds have a keel-shaped
breastbone that anchors large
pectoral muscles. These give
the bird enough strength to fly.

Humerus

Pectoral muscles

Keeled sternum

DIGESTIVE SYSTEM

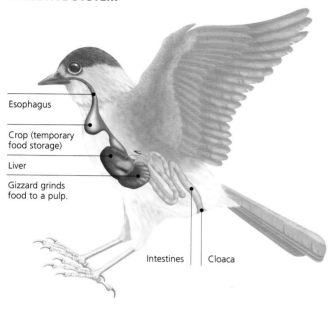

Esophagus

Crop (temporary food storage)

Liver

Gizzard grinds food to a pulp.

Intestines

Cloaca

LEG MUSCLES

All birds have powerful leg muscles in the top of the leg, near their center of gravity. These are connected to the toes by long tendons.

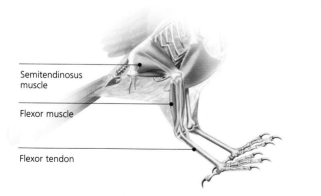

Semitendinosus muscle

Flexor muscle

Flexor tendon

HOLLOW BONES

Most bones in a bird's body are thin-walled and hollow, with struts and braces providing maximum strength for minimum weight.

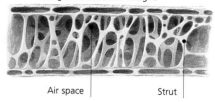

Air space

Strut

LUNG CROSS-SECTION

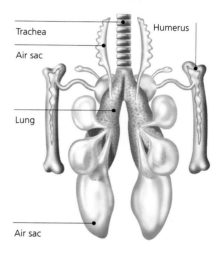

Trachea

Humerus

Air sac

Lung

Air sac

HEART CROSS-SECTION

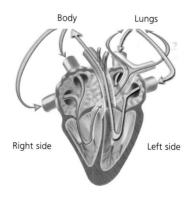

Body

Lungs

Right side

Left side

Flightless Birds

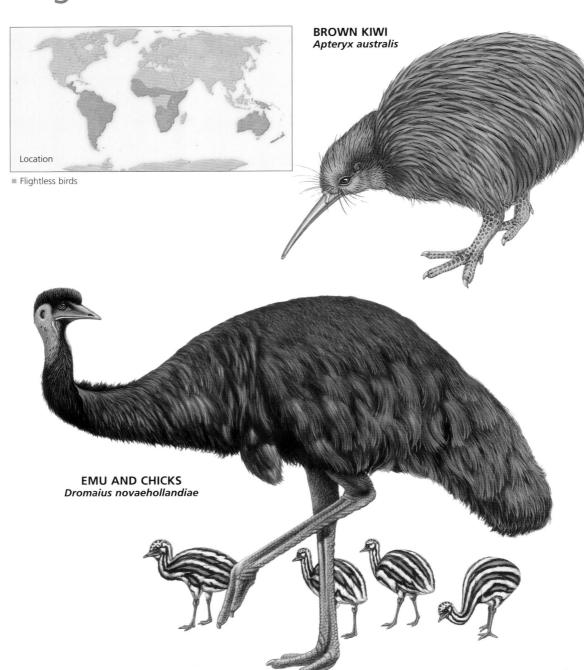

Location

■ Flightless birds

BROWN KIWI
Apteryx australis

EMU AND CHICKS
Dromaius novaehollandiae

Female great tinamou head

Emu at rest

Emu egg

Great tinamou

Greater rhea

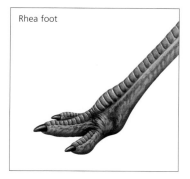
Rhea foot

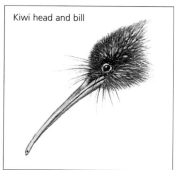
Kiwi using claws to scratch for food

Ostrich tail feather

Ostrich

Kiwi head and bill

Cassowary foot

Southern cassowary

Penguins

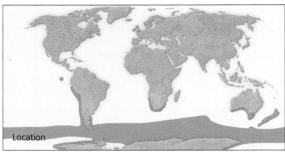

Location

■ Penguins

EMPEROR PENGUIN ANATOMY

Feathers

Foot

| Scaly, oily tips | Bend at the base | Fluffy down | Ankle feathers | Long toenails to grip the ice |

EMPEROR PENGUINS SWIMMING
Emperor penguins are swift, agile swimmers, and can swim up to 20 miles (32 km) per hour. They dive through a hole in the ice and use their stiff, flipperlike wings to chase their prey.

FJORDLAND PENGUIN
Eudyptes pachyrhynchus

ROCKHOPPER PENGUIN
Eudyptes chrysocome

MAGELLANIC PENGUIN
Spheniscus magellanicus

FAIRY PENGUIN
Eudyptula minor

EMPEROR PENGUIN
Aptenodytes forsteri

CHINSTRAP PENGUIN
Pygoscelis antarctica

YELLOW-EYED PENGUIN
Megadyptes antipodes

Grebes and Divers

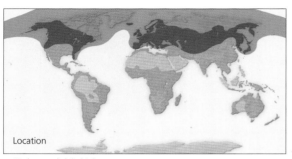

Location

- Grebes and dabchicks
- Divers (loons)
- Grebes, dabchicks and divers

GREAT CRESTED GREBE
Podiceps cristatus

RED-NECKED GREBE
Podiceps grisegena

PIED-BILLED GREBE WITH CHICKS
Podilymbus podiceps

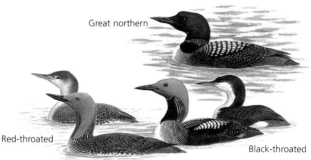

Great northern

Red-throated

Black-throated

GROUP OF DIVERS (LOONS)
Gavia sp.

COMMON LOON
Gavia immer

HORNED GREBE
Podiceps auritus

EARED GREBE
Podiceps nigricollis

WESTERN GREBES 'RUSHING'

Rushing involves two or more grebes running upright across the water as part of a ritualistic courtship display.

Lakeside vegetation

Necks are arched with head slightly bowed.

Wings are drawn back and flexed.

Males use this display to defend their territory.

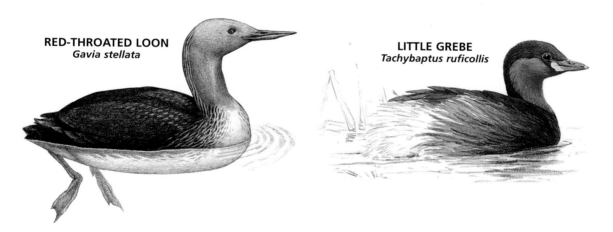

RED-THROATED LOON
Gavia stellata

LITTLE GREBE
Tachybaptus ruficollis

Pacific (left) and common loon heads

Western (top) and Clark's grebe heads

Yellow-billed loon head and neck

Albatrosses and Petrels

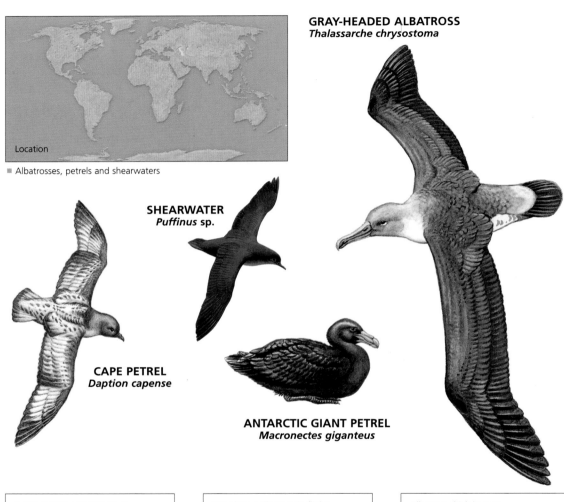

Location

■ Albatrosses, petrels and shearwaters

GRAY-HEADED ALBATROSS
Thalassarche chrysostoma

SHEARWATER
Puffinus sp.

CAPE PETREL
Daption capense

ANTARCTIC GIANT PETREL
Macronectes giganteus

Royal albatross mating ritual

Wandering albatross in flight

Albatross chick in nest

SHORT-TAILED SHEARWATER MIGRATION

June–August
Shearwaters spend the non-breeding season in the North Pacific Ocean.

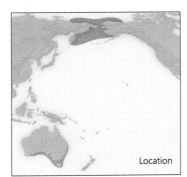

April–May
Chick reaches adult weight and grows flight feathers, ready to leave the nest.

February–March
The young chick is fed every three days as its parents alternate between care and hunting.

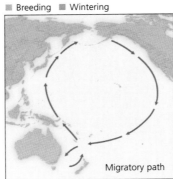

Location

■ Breeding ■ Wintering

Migratory path

September
It takes several weeks for the birds migrating south to reach their nesting colonies, along the coast of Australia.

October–November
After courting, shearwaters establish pair bonds for the year, and repair their nest site.

December–January
Both parents incubate the egg, laid late November, until it hatches in January.

Pelicans

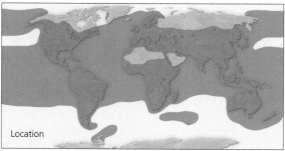

Location

■ Pelicans

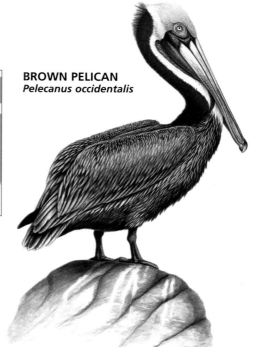

BROWN PELICAN
Pelecanus occidentalis

Shag with summer plumage

British cormorant spreading wings

Blue-footed booby on rock

Magnificent frigatebird with throat sac

Gannet at rest

Australian pelican

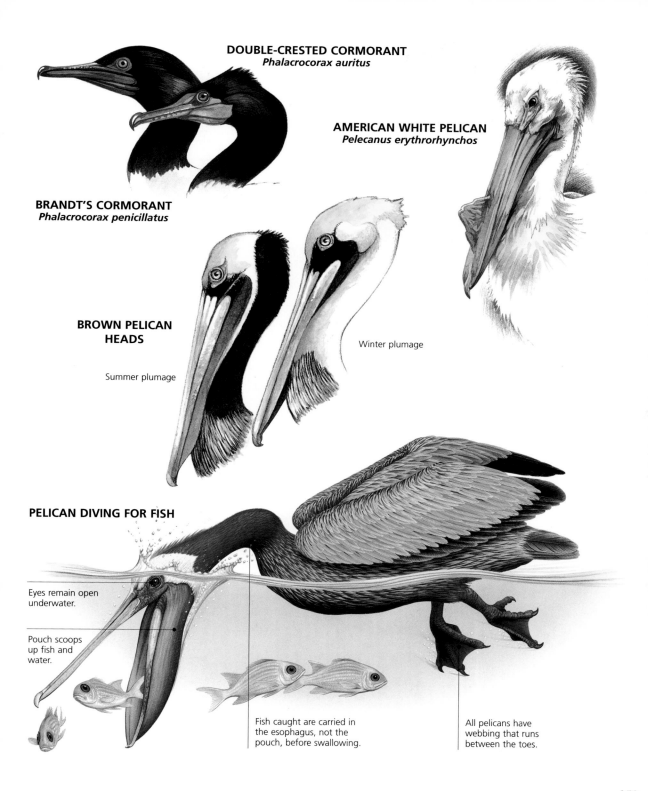

DOUBLE-CRESTED CORMORANT
Phalacrocorax auritus

AMERICAN WHITE PELICAN
Pelecanus erythrorhynchos

BRANDT'S CORMORANT
Phalacrocorax penicillatus

BROWN PELICAN HEADS

Summer plumage

Winter plumage

PELICAN DIVING FOR FISH

Eyes remain open underwater.

Pouch scoops up fish and water.

Fish caught are carried in the esophagus, not the pouch, before swallowing.

All pelicans have webbing that runs between the toes.

Herons

Location

■ Herons

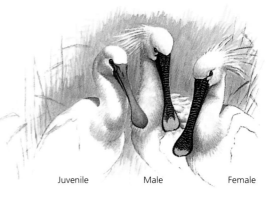

Juvenile   Male   Female

EURASIAN SPOONBILLS
Platalea leucorodia

Adult

Juvenile

BLACK-CROWNED NIGHT HERONS
Nycticorax nycticorax

GREAT BITTERN
Botaurus stellaris

**SHOEBILL
WITH YOUNG**
Balaeniceps rex

LITTLE EGRET
Egretta garzetta

**GREAT WHITE
EGRET**
Egretta alba

Yellow-crowned night heron head

Roseate spoonbill spreading wings

Roseate spoonbill at rest

Juvenile little blue heron

Great blue heron in flight

African spoonbill head

Black stork wading

Great blue heron head

Scarlet ibis and skeleton

White stork in flight

Great bittern camouflaged in reeds

Great (left) and little bitterns in flight

Herons

GREEN HERONS FISHING
The green herons of Florida, U.S.A., have learnt that dropping breadcrumbs into the water attracts fish. The next generation are then taught the same tactic.

GRAY HERON
Ardea cinerea

STRIATED HERON
Butorides striatus

GREEN HERON
Butorides virescens

WALDRAPP
Geronticus eremita

LITTLE BLUE HERON
Egretta caerulea

CATTLE EGRET
Bubulcus ibis

PURPLE HERON
Ardea purpurea

GOLIATH HERON
Ardea goliath

PAINTED STORK
Mycteria leucocephala

ROSEATE SPOONBILL
Ajaia ajaja

257

Flamingos

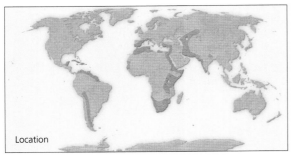

Location

■ Flamingos

LESSER FLAMINGO FEEDING

Long, flexible neck bends to the water while the flamingo remains upright.

Upper jaw contains a row of slits, called lamellae, that act like a sieve.

Tongue and lower jaw strain food from the water through the lamellae.

LESSER FLAMINGO
Phoenicopterus minor

PUNA FLAMINGO
Phoenicopterus jamesi

Flamingo at rest

Flamingo feeding

Greater flamingos

GREATER FLAMINGO COLONY

Flamingos fly with their neck stretched out and long legs trailing behind.

Flamingos are highly social birds that live and breed in large lakeside colonies.

Their dramatic pink coloring is thought to come from certain algae that they eat.

A hooked bill scoops up water containing tiny animals and algae.

Waterfowl

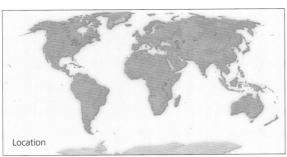

Location

■ Waterfowl

COMMON MOORHENS
Gallinula chloropus

BARNACLE GEESE MIGRATING
Branta leucopsis
Barnacle geese fly in a characteristic
v-shaped formation when migrating
to and from Greenland and Europe.

Lead bird flies at the head of
the v-shape, breaking up air
currents for those that follow.

Migrating birds spend a short
time as lead bird, before
dropping back into the flock.

Canada goose with goslings

Velvet scoters

Gaggle of Perigord geese

Mallard foot showing webbed toes

Mallard in flight

Mallard diving for fish

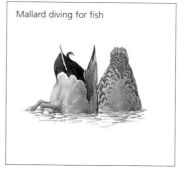

Ruddy duck diving

Dark- and pale-bellied Brant geese

Black swan swimming

Northen shoveler with spatulate bill

Harlequin ducks swimming

Whooper and Bewick's swans

Waterfowl

RED-BREASTED GOOSE
Branta ruficollis

KING EIDER
Somateria spectabilis

WOOD DUCK
Aix sponsa

NORTHERN SCREAMER
Chauna chavaria

GRAYLAG GEESE
Anser anser

COMMON GOLDENEYES
Bucephala clangula

Male

Female

SNOW GOOSE
Chen caerulescens

AMERICAN BLACK DUCK
Anas rubripes

CINNAMON TEAL
Anas cyanoptera

Adult

Juvenile

MUTE SWANS
Cygnus olor

FEMALE COMMON EIDERS WITH DUCKLINGS
Somateria mollissima

263

Birds of Prey

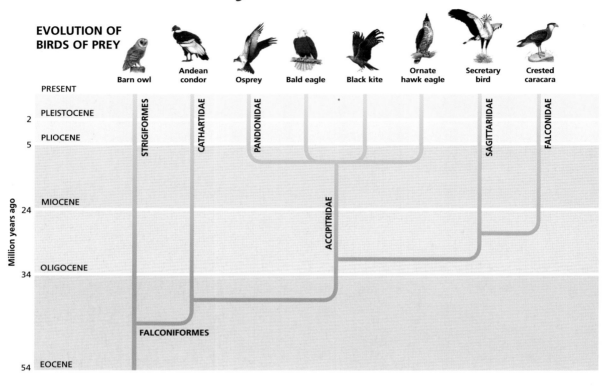

**EVOLUTION OF
BIRDS OF PREY**

Barn owl   Andean condor   Osprey   Bald eagle   Black kite   Ornate hawk eagle   Secretary bird   Crested caracara

PRESENT

PLEISTOCENE

2

PLIOCENE

5

STRIGIFORMES   CATHARTIDAE   PANDIONIDAE

MIOCENE

24

ACCIPITRIDAE

SAGITTARIIDAE   FALCONIDAE

OLIGOCENE

34

Million years ago

FALCONIFORMES

EOCENE

54

WING SPAN COMPARISON

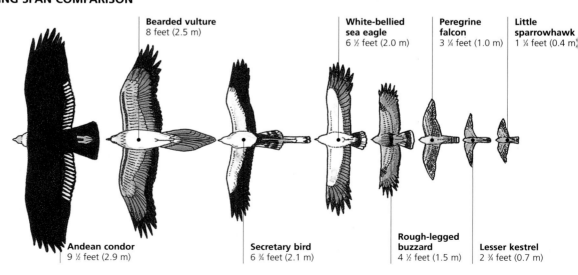

Bearded vulture
8 feet (2.5 m)

**White-bellied
sea eagle**
6 ½ feet (2.0 m)

**Peregrine
falcon**
3 ¼ feet (1.0 m)

**Little
sparrowhawk**
1 ¼ feet (0.4 m)

Andean condor
9 ½ feet (2.9 m)

Secretary bird
6 ¾ feet (2.1 m)

**Rough-legged
buzzard**
4 ½ feet (1.5 m)

Lesser kestrel
2 ¾ feet (0.7 m)

RAPTOR ANATOMY

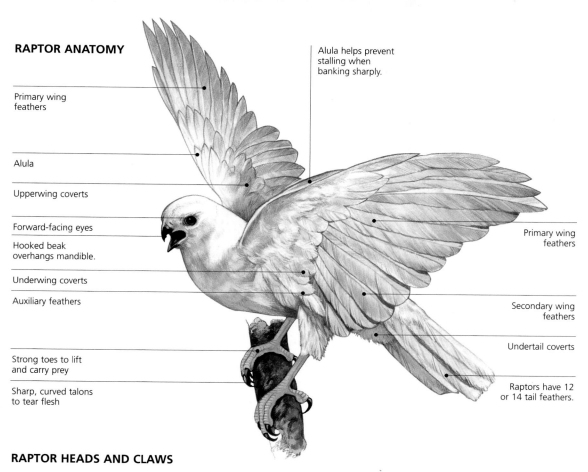

Primary wing feathers

Alula helps prevent stalling when banking sharply.

Alula

Upperwing coverts

Forward-facing eyes

Hooked beak overhangs mandible.

Underwing coverts

Auxiliary feathers

Primary wing feathers

Secondary wing feathers

Undertail coverts

Strong toes to lift and carry prey

Sharp, curved talons to tear flesh

Raptors have 12 or 14 tail feathers.

RAPTOR HEADS AND CLAWS

Osprey

White-backed vulture

Eurasian sparrowhawk

Harpy eagle

Osprey: fish eater

Vuture: carrion eater

Sparrowhawk: bird eater

Harpy eagle: mammal eater

Eagles, Kites and Hawks

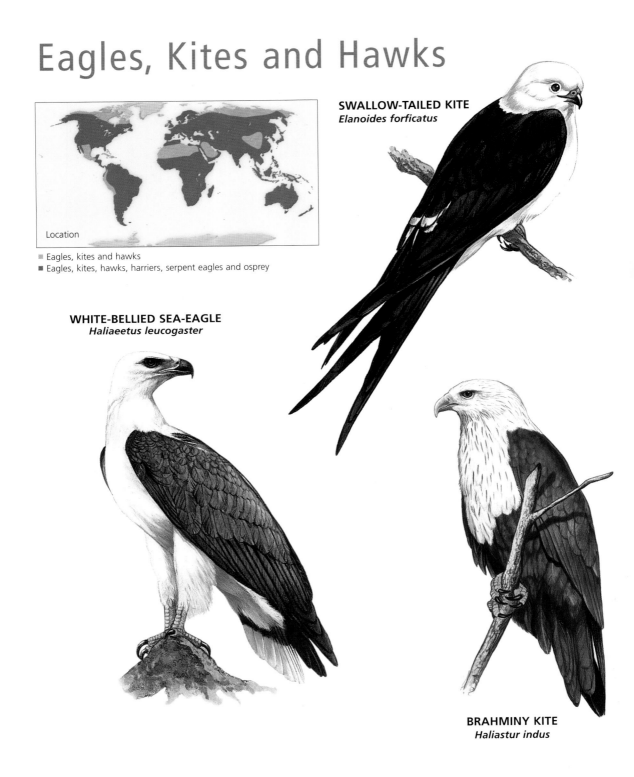

Location

- Eagles, kites and hawks
- Eagles, kites, hawks, harriers, serpent eagles and osprey

SWALLOW-TAILED KITE
Elanoides forficatus

WHITE-BELLIED SEA-EAGLE
Haliaeetus leucogaster

BRAHMINY KITE
Haliastur indus

HOOK-BILLED KITE
Chondrohierax uncinatus

SNAIL KITE
Rostrhamus sociabilis

CRESTED SERPENT-EAGLE
Spilornis cheela

BLACK-SHOULDERED KITE
Elanus caeruleus

GREAT PHILIPPINE EAGLE
Pithecophaga jefferyi

Eagles, Kites and Hawks

PEARL KITE
Gampsonyx swainsonii

SLATE-COLORED HAWK
Leucopternis schistacea

ROUGH-LEGGED HAWK
Buteo lagopus

OSPREY
Pandion haliaetus

ORNATE HAWK EAGLE
Spizaetus ornatus

AFRICAN HARRIER-HAWK
Polyboroides typus

PACIFIC BAZA
Aviceda subcristata

EASTERN CHANTING GOSHAWK
Melierax poliopterus

BATELEUR
Terathopius ecaudatus

269

Falcons

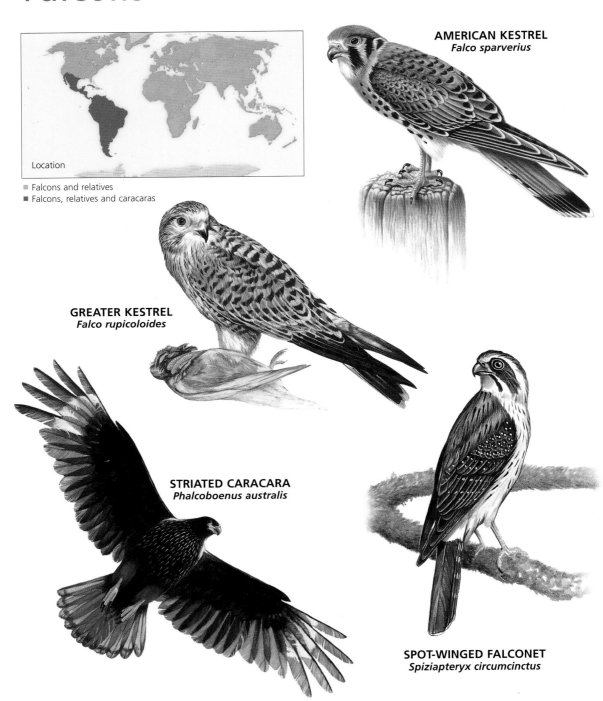

Location
- Falcons and relatives
- Falcons, relatives and caracaras

AMERICAN KESTREL
Falco sparverius

GREATER KESTREL
Falco rupicoloides

STRIATED CARACARA
Phalcoboenus australis

SPOT-WINGED FALCONET
Spiziapteryx circumcinctus

Lined forest falcon

Peregrine falcon hunting pigeon

American kestrel displaying plumage

PEREGRINE FALCON
Falco peregrinus

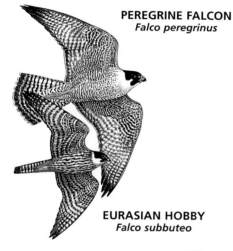

PYGMY FALCON
Polihierax semitorquatus

EURASIAN HOBBY
Falco subbuteo

PEREGRINE FALCONS MATING
The male falcon alights on the female's lowered back, and balances by flapping furiously. He lowers his tail under hers so that their cloacas meet and sperm is exchanged.

CRESTED CARACARA
Caracara plancus

Secretary Bird and Vultures

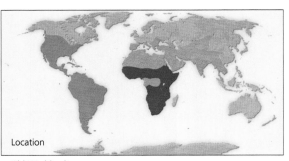

Location

- Old World vultures
- New World vultures
- Old World vultures and secretary bird

ANDEAN CONDOR
Vultur gryphus

PALM-NUT VULTURE
Gypohierax angolensis

SECRETARY BIRD
Sagittarius serpentarius

LAMMERGEIER
Gypaetus barbatus

Lappet-faced vulture spreading wings

Griffon vulture on rock

Secretary bird grabbing snake with feet

Lappet-faced vulture

Juvenile lammergeier

Black vulture head

Eurasian griffon

Juvenile (left) and adult turkey vultures

Turkey vulture in flight

Gamebirds

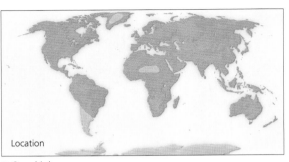

Location

■ Gamebirds

MALEO
Macrocephalon maleo

MALLEEFOWL NEST

Sand layer makes the nest warmer or cooler.

Leaf litter layer slowly rots to generate heat.

Eggs incubate at the constant temperature of about 91° F (33° C).

Malleefowls eat herbs in winter, fruits and seeds in summer.

They do not drink water, obtaining all necessary liquids from their solid diet.

Male tests temperature inside the nest with his mouth and tongue.

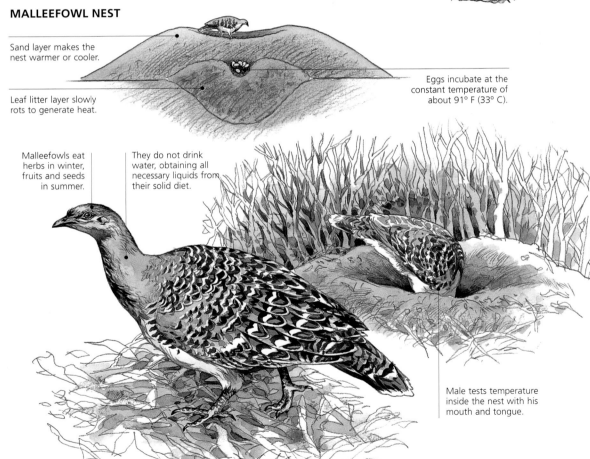

Chicken egg

Wild turkey displaying plumage

Baby fowl and egg

Gray partridges preparing to dust bath

Ptarmigan with winter plumage

Ptarmigan with summer plumage

Ruffed grouse displaying tail feathers

Ring-necked pheasant skeleton

Ring-necked pheasant

Temminck's tragopan

Satyr tragopan

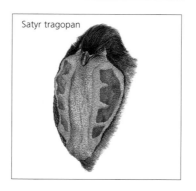

Capercaillie displaying tail feathers

Gamebirds

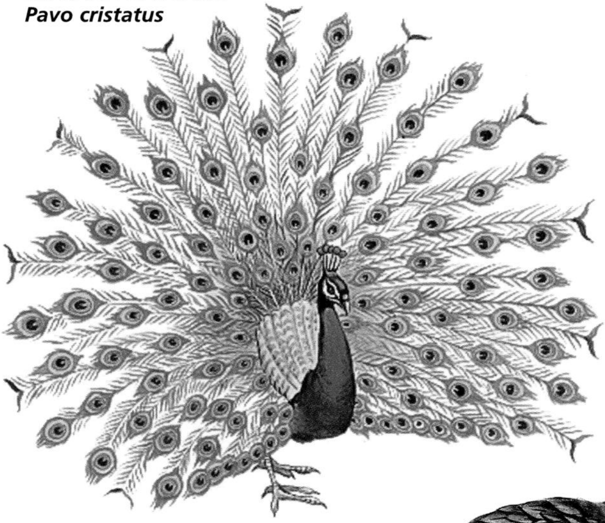

Gamebirds

BIRDS

INDIAN PEAFOWL
Pavo cristatus

HOATZIN
Opisthocomus hoazin

WHITE-CRESTED GUAN
Penelope pileata

GREATER PRAIRIE-CHICKEN
Tympanuchus cupido

CALIFORNIA QUAIL
Callipepla californica

276

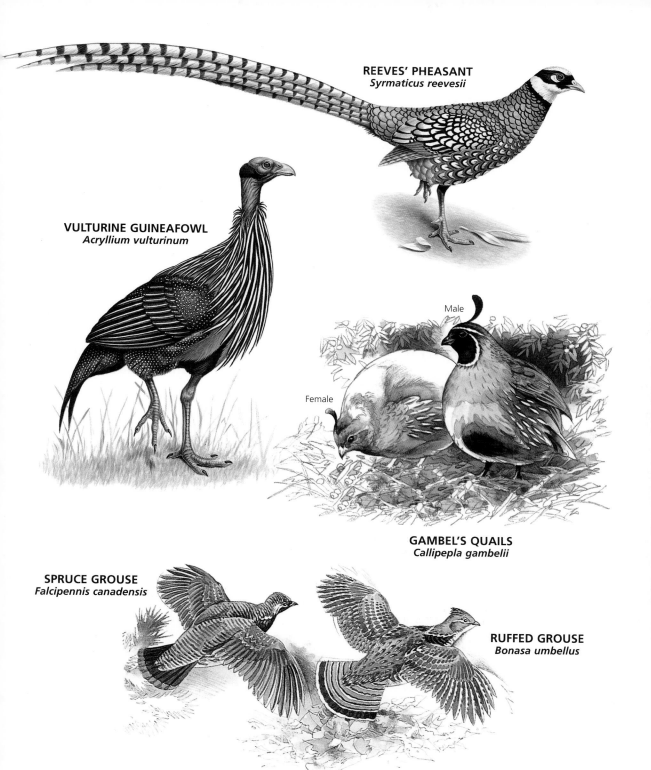

REEVES' PHEASANT
Syrmaticus reevesii

VULTURINE GUINEAFOWL
Acryllium vulturinum

Male

Female

GAMBEL'S QUAILS
Callipepla gambelii

SPRUCE GROUSE
Falcipennis canadensis

RUFFED GROUSE
Bonasa umbellus

277

Cranes

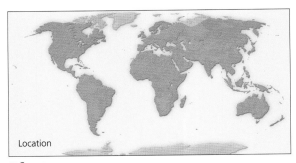

Location

▪ Cranes

LIMPKIN
Aramus guarauna

RED-NECKED CRAKE
Rallina tricolor

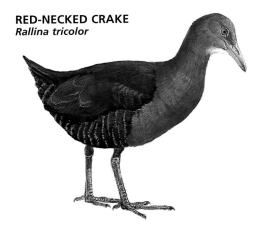

WOODFORD'S RAIL
Nesoclopeus woodfordi

SIBERIAN CRANE MATING DANCE

Male initiates mating dance.   He performs rapid bows before stretching his wings and calling.   The female responds, and is imitated.

SUNBITTERN
Eurypyga helias

BLACK BUSTARD
Eupodotis afra

PURPLE GALLINULE
Porphyrula martinica

BLACK CROWNED-CRANE
Balearica pavonina

279

Cranes

TAKAHE
Porphyrio mantelli

COMMON MOORHEN
Gallinula chloropus

BUFF-BANDED RAIL
Gallirallus philippensis

WEKA
Gallirallus australis

EURASIAN COOT
Fulica atra

Juvenile (left) and adult moorhens

Kagu

Sandhill cranes migrating

Water rail chicks

Water rail

Juvenile rail skeleton

Red-crowned crane

White-quilled bustard

Plains-wanderer

Okinawa rail

Common cranes displaying

Houbara bustard

281

Waders and Shorebirds

WADER AND SHOREBIRD BEHAVIOR

Oystercatchers

Individual oystercatchers specialize in catching specific foods, such as worms, crabs, oysters or mussels.

Sandwich terns

Male and female sandwich terns share small fishes as part of their courtship display during the breeding season.

Horned puffins feeding

Horned puffins dive deeply into the ocean to catch small fishes, which are held in their mouth by backward-facing grooves until they are ready to eat them.

Avocet with chicks

Avocets feed by sweeping their long, upturned bill through the mud while slowly walking forward.

Turnstones

As their name suggests, turnstones walk purposefully over the mudflats, turning over stones, seaweed or wood with their beaks in search of small crustaceans and worms.

Spotted sandpiper in summer plumage

Willets in winter (left) and breeding plumage

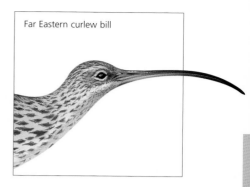
Far Eastern curlew bill

Arctic tern in flight

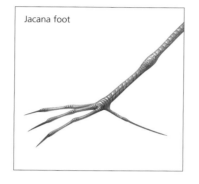
Jacana foot

Skua spreading wings

Godwit (top), dowitcher and curlew bills

Common murre egg

Golden plovers in flight

Black-headed gull facial plumage

Winter

Summer

Ruddy turnstone

Crab plovers digging nest holes

Waders

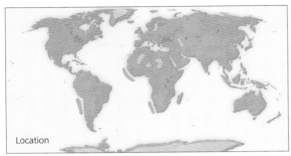

Location

■ Waders

COMMON SNIPE
Gallinago gallinago

BANDED LAPWING
Vanellus tricolor

BLACK-BELLIED PLOVER
Pluvialis squatarola

Summer plumage

Winter plumage

Winter plumage

KILLDEER
Charadrius vociferus
The killdeer acts out a 'broken wing' display to distract and draw predators from the nest.

SPOTTED SANDPIPER
Actitis macularia

AMERICAN GOLDEN-PLOVER
Pluvialis dominica

BLACK-TAILED GODWIT
Limosa limosa

DUNLINS
Calidris alpina

Adult summer
plumage

Adult winter
plumage

Juvenile

BLACK SKIMMER
Rynchops niger

Shorebirds and Seabirds

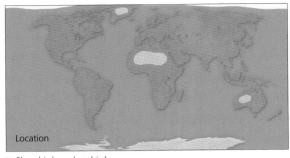

Location

■ Shorebirds and seabirds

GULL WING CUTAWAY

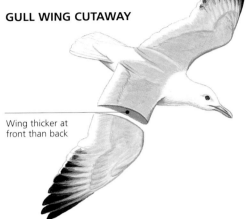

Wing thicker at
front than back

SHOREBIRD HABITAT

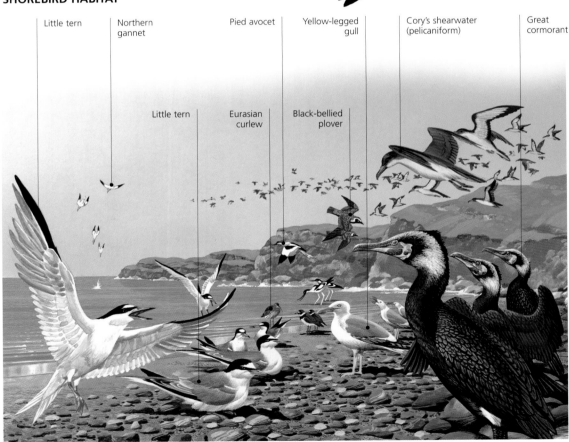

Little tern

Northern
gannet

Pied avocet

Yellow-legged
gull

Cory's shearwater
(pelicaniform)

Great
cormorant

Little tern

Eurasian
curlew

Black-bellied
plover

RING-BILLED GULL
Larus delawarensis

ARCTIC TERN
Sterna paradisaea

CASPIAN TERNS
Terna caspia

Adult summer plumage

Juvenile

ATLANTIC PUFFIN
Fratercula arctica

YELLOW-LEGGED GULL
Larus cachinnans

MEW GULL
Larus canus

HERRING GULL
Larus argentatus

Juvenile herring gull

Common (left) and black guillemots

Common tern eggs hidden in rocks

Royal (top) and Caspian tern heads

Pigeons and Sandgrouse

Location

■ Pigeons and sandgrouse

SUPERB FRUIT-DOVE
Ptilinopus superbus

WHITE-WINGED DOVE
Zenaida asiatica

EURASIAN TURTLE DOVE
Streptopelia turtur

PIN-TAILED SANDGROUSE
Pterocles alchata

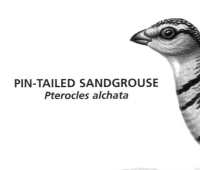

ROCK DOVES
Columba livia

NICOBAR PIGEON
Caloenas nicobarica

Lichtenstein's sandgrouse

Male and female pin-tailed sandgrouse

Rock dove face and bill

Pigeon in flight

Crowned pigeon

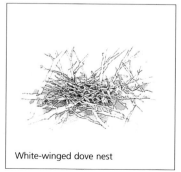

White-winged dove nest

Parrots

PARROT SKULL

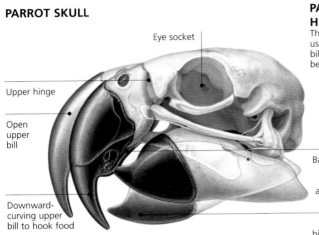

Eye socket

Upper hinge

Open upper bill

Downward-curving upper bill to hook food

Base of upper bill used to crack seeds against curve of lower bill

Open lower bill (mandible)

PALM COCKATOO HEAD AND BILL

The palm cockatoo uses its broad, curved bill to crush fruit, berries and seeds.

TYPES OF PARROT FEATHERS

Macaw body feather

Red-fan parrot facial feathers

Cockatiel flight feather

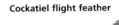

SCARLET MACAWS IN FLIGHT

Scarlet macaws (*Ara macao*) live in the rainforest canopies of Central and South America. Their brilliant, primary-colored plumage has made them popular with poachers and their numbers are now seriously depleted.

Australian night parrot

Kakapo

Black cockatoo

Kea

Scarlet macaw

Sulfur-crested cockatoo

Eastern rosella in flight

Gray parrot

Black-capped lory drinking nectar

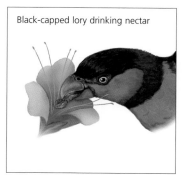

Budgerigars

Lorikeet foot

Rosy-faced lovebird on branch

Parrots

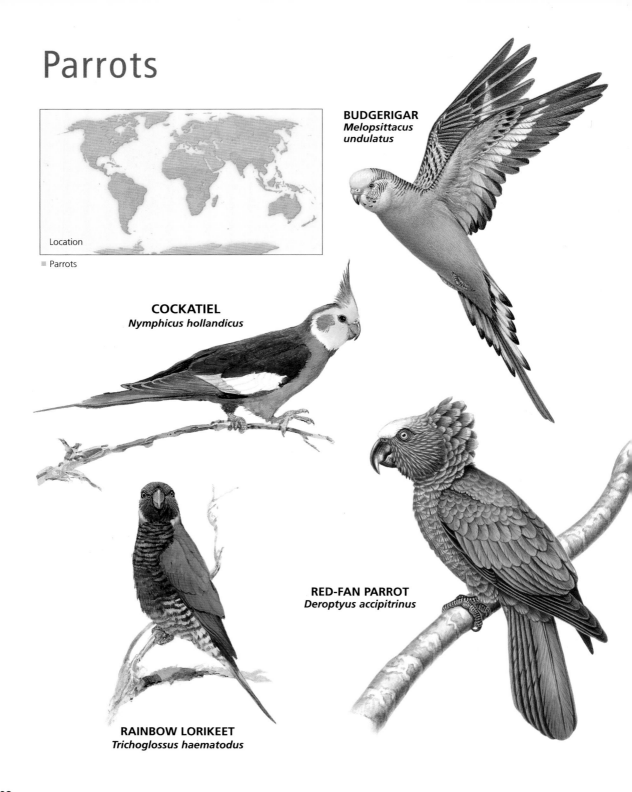

Location

■ Parrots

BUDGERIGAR
Melopsittacus undulatus

COCKATIEL
Nymphicus hollandicus

RED-FAN PARROT
Deroptyus accipitrinus

RAINBOW LORIKEET
Trichoglossus haematodus

BLACK-CAPPED LORY
Lorius lory

EASTERN ROSELLA
Platycercus eximius

PESQUET'S PARROT
Psittrichas fulgidus

**SALMON-CRESTED
COCKATOO**
Cacatua moluccensis

293

Cuckoos and Turacos

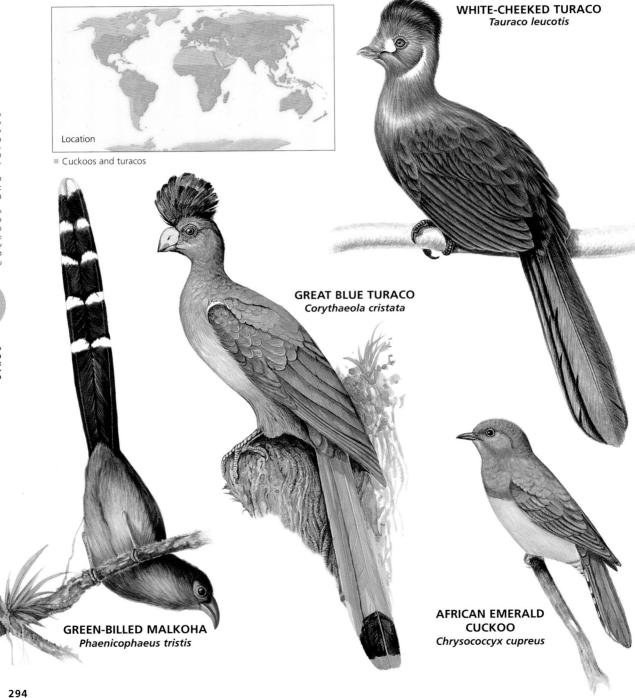

Location

■ Cuckoos and turacos

WHITE-CHEEKED TURACO
Tauraco leucotis

GREAT BLUE TURACO
Corythaeola cristata

GREEN-BILLED MALKOHA
Phaenicophaeus tristis

AFRICAN EMERALD CUCKOO
Chrysococcyx cupreus

CHANNEL-BILLED CUCKOO
Scythrops novaehollandiae

GREATER ROADRUNNER
Geococcyx californianus

COMMON CUCKOO
Cuculus canorus

Cuckoo nest

Greater roadrunner
in defensive posture

Violet turaco on branch

Owls

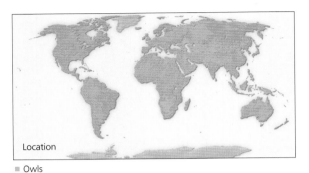
Location
- Owls

OWL HUNTING

Glides silently toward prey

Raises wings and extends talons

Grabs prey with talons, dropping wings for balance

OWL SKULL
Some owl species have asymmetrical ears to provide greater accuracy in locating prey.

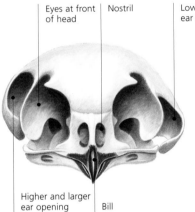

Eyes at front of head

Nostril

Lower and smaller ear opening

Higher and larger ear opening

Bill

OWL VISION

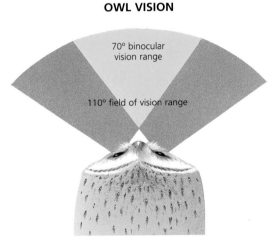

70° binocular vision range

110° field of vision range

Barn owl talons

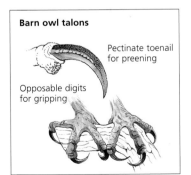

Pectinate toenail for preening

Opposable digits for gripping

Flight feathers are serrated to reduce noise.

Barn owl head and shoulders

SPOTTED OWL
Strix occidentalis

BARN OWL
Tyto alba

LITTLE OWL
Athene noctua

**GREAT HORNED OWL
WITH YOUNG**
Bubo virginianus

Snowy owl

Barn owl chicks

Barred owl chicks in nest

Nightjars and Frogmouths

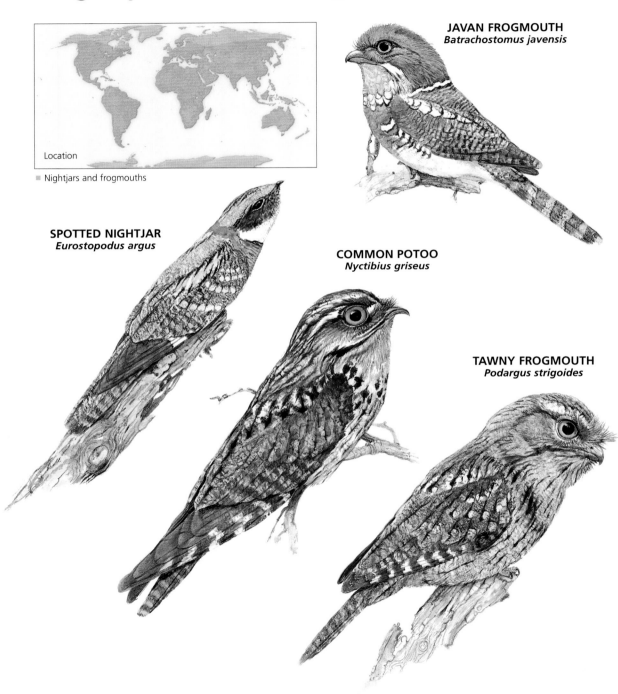

Location

■ Nightjars and frogmouths

JAVAN FROGMOUTH
Batrachostomus javensis

SPOTTED NIGHTJAR
Eurostopodus argus

COMMON POTOO
Nyctibius griseus

TAWNY FROGMOUTH
Podargus strigoides

Nightjars and Frogmouths

BIRDS

OWLETTE NIGHTJAR
Aegotheles sp.

PAURAQUE
Nyctidromus albicollis

INDIAN NIGHTJAR
Caprimulgus asiaticus

OILBIRD
Steatornis caripensis

Oilbird showing back and tail plumage

Common nighthawk catching prey

Nightjar hunting

Hummingbirds and Swifts

PURPLE-BACKED THORNBILL HOVERING

The purple-backed thornbill's wings beat rapidly in a number of directions, rotating around the flexible shoulder joint to enable the bird to hover.

Up and down

Forward and below head

Figure-eight around shoulder joint

Above and behind head

HUMMINGBIRD FEEDING

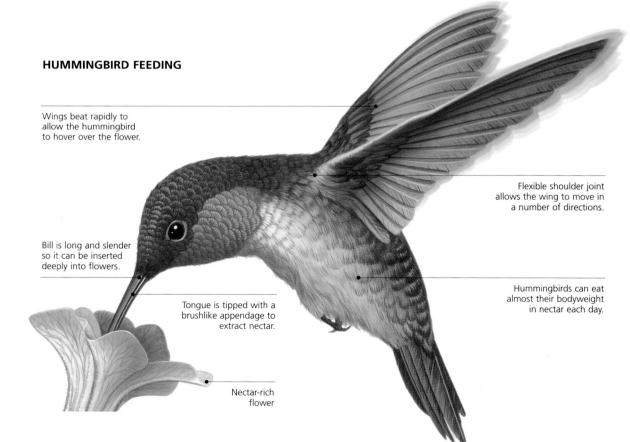

Wings beat rapidly to allow the hummingbird to hover over the flower.

Flexible shoulder joint allows the wing to move in a number of directions.

Bill is long and slender so it can be inserted deeply into flowers.

Hummingbirds can eat almost their bodyweight in nectar each day.

Tongue is tipped with a brushlike appendage to extract nectar.

Nectar-rich flower

Hummingbird hovering

White-throated needletail gliding

Rufous hummingbird in flight

Costa's hummingbird feeding

Swifts in flight over rooftops

Black-chinned hummingbird head

Anna's hummingbird feeding

Anna's hummingbird egg

Black-chinned hummingbird hovering

Male ruby-throated hummingbird head

Ruby-throated hummingbird on nest

Chimney (top) and Vaux's swifts in flight

Hummingbirds and Swifts

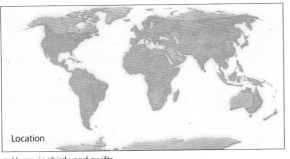

Location

■ Hummingbirds and swifts

LITTLE SWIFT
Apus affinis

RUFOUS HUMMINGBIRD
Selasphorus rufus

**BLACK-CHINNED
HUMMINGBIRD**
Archilochus alexandri

RUBY-THROATED HUMMINGBIRD
Archilochus colubris

GRAY-RUMPED TREESWIFT
Hemiprocne longipennis

ANNA'S HUMMINGBIRD
Calypte anna

COSTA'S HUMMINGBIRD
Calypte costae

ALPINE SWIFT
Tachymarptis melba

CALLIOPE HUMMINGBIRD
Stellula calliope

BROAD-TAILED HUMMINGBIRD
Selasphorus platycercus

BLUE-THROATED HUMMINGBIRD
Lampornis clemenciae

Mousebirds and Trogons

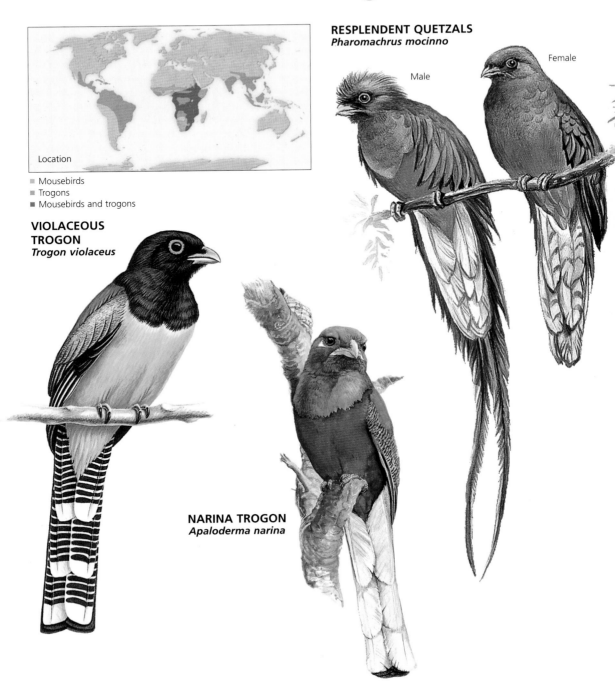

Location

- Mousebirds
- Trogons
- Mousebirds and trogons

RESPLENDENT QUETZALS
Pharomachrus mocinno

Male

Female

VIOLACEOUS TROGON
Trogon violaceus

NARINA TROGON
Apaloderma narina

SPECKLED MOUSEBIRD
Colius striatus

RED-HEADED TROGON
Harpactes erythrocephalus

WHITE-TAILED TROGON
Trogon viridis

Scarlet-rumped trogon

Diard's trogon

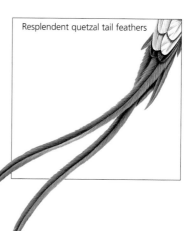

Resplendent quetzal tail feathers

305

Kingfishers

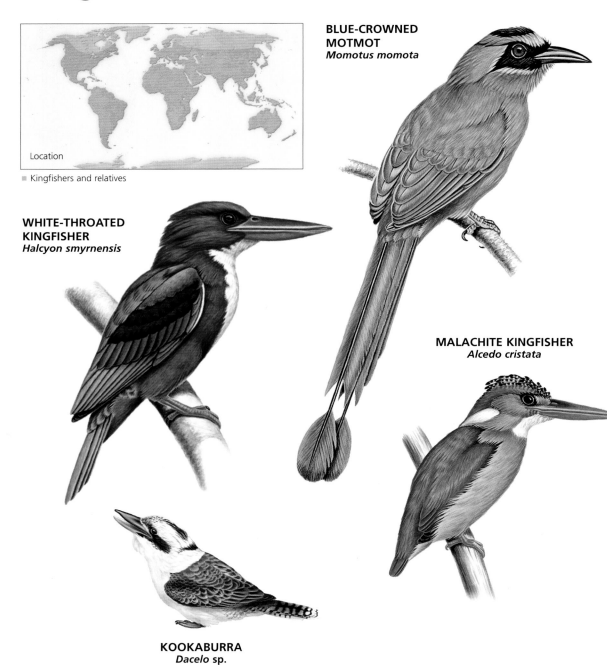

■ Kingfishers and relatives

BLUE-CROWNED MOTMOT
Momotus momota

WHITE-THROATED KINGFISHER
Halcyon smyrnensis

MALACHITE KINGFISHER
Alcedo cristata

KOOKABURRA
Dacelo sp.

Cuban tody on branch

Rufous motmot

Eurasian kingfisher searching for prey

Banded kookaburra

Bee-eater in excavated nest cavity

Long-tailed ground-roller

Kookaburras on branch

Colorful kingfisher

Rufous-headed ground-roller

Hoopoe with young hidden in nest

Hoopoe with crest flattened

Hoopoe with crest erect

Kingfishers

**NORTHERN CARMINE
BEE-EATER**
Merops nubicus

**RAINBOW
BEE-EATER**
Merops ornatus

**BELTED KINGFISHER
FISHING**

Female flies in search
of prey.

She spots a fish.

She swoops down and
dives into the water.

DOLLARBIRD
Eurystomus orientalis

The nest is a tunnel dug into a riverbank,
above the waterline to avoid flooding.

Catching the fish in her bill,
she returns to the nest.

SPANGLED KOOKABURRA
Dacelo tyro

LILAC-BREASTED ROLLER
Coracias caudata

YELLOW-BILLED HORNBILL
Tockus sp.

Woodpeckers and Toucans

Location

■ Woodpeckers, barbets and toucans

TOCO TOUCAN
Ramphastos toco

Fruit-eaters, toucans use their bill to pick up fruit, before tossing it back into their mouth to swallow.

The toco toucan's large bill is hollow and light, strengthened inside by a honeycomblike structure.

Toes positioned two in front, two behind, allow the toco toucan to grasp hold of branches.

Toco toucans sleep by resting their long beak over their back and folding their tail back over their head.

PILEATED WOODPECKER HEAD
Dryocopus pileatus

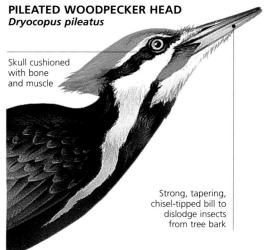

Skull cushioned with bone and muscle

Strong, tapering, chisel-tipped bill to dislodge insects from tree bark

WOODPECKER TONGUE

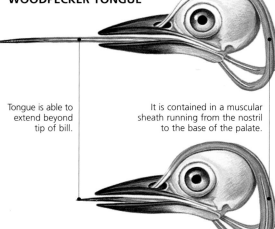

Tongue is able to extend beyond tip of bill.

It is contained in a muscular sheath running from the nostril to the base of the palate.

Great (left)
and lesser spotted woodpeckers

Maroon woodpecker on branch

Chestnut-capped puffbird

Yellow-rumped honeyguide

Green woodpecker digging in anthill

Red-bellied woodpecker

Acorn woodpecker storing acorn

Rufous piculet

Hairy woodpecker tapping tree bark

Acorn woodpecker in flight

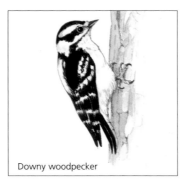
Downy woodpecker

Hairy and downy woodpecker tails

Woodpeckers and Toucans

NORTHERN FLICKER SUBSPECIES

Gilded flicker

Red-shafted flicker

Yellow-shafted flicker

NORTHERN FLICKER
Colaptes auratus

KEEL-BILLED TOUCAN
Ramphastos sulfuratos

RED-AND-YELLOW BARBET
Trachyphonus erythrocephalus

CURL-CRESTED ARACARI
Pteroglossus beauharnaesii

RUFOUS-TAILED JACAMAR
Galbula ruficauda

EMERALD TOUCANET
Aulacorhynchus prasinus

BLACK-RUMPED FLAMEBACK
Dinopium benghalense

DOUBLE-TOOTHED BARBET
Lybius bidentatus

Passerines

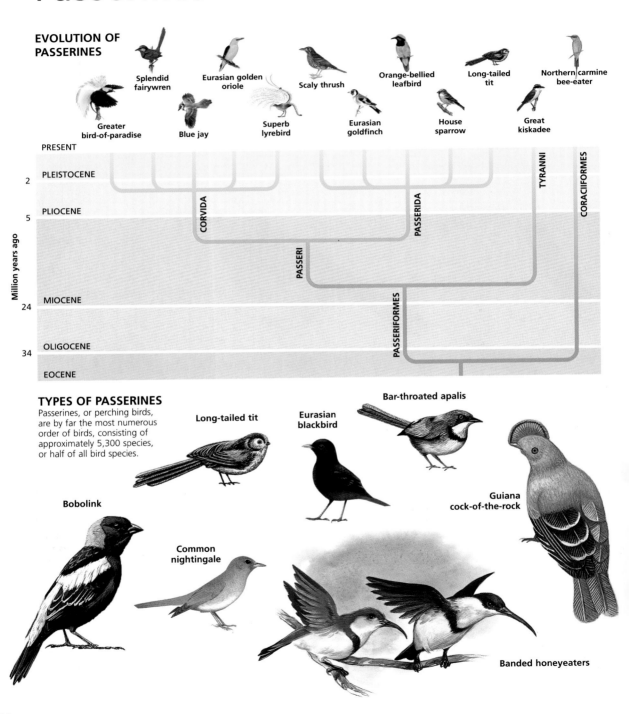

EVOLUTION OF PASSERINES

Splendid fairywren

Eurasian golden oriole

Scaly thrush

Orange-bellied leafbird

Long-tailed tit

Northern carmine bee-eater

Greater bird-of-paradise

Blue jay

Superb lyrebird

Eurasian goldfinch

House sparrow

Great kiskadee

PRESENT

PLEISTOCENE

2

PLIOCENE

5

MIOCENE

24

OLIGOCENE

34

EOCENE

Million years ago

CORVIDA

PASSERI

PASSERIDA

PASSERIFORMES

TYRANNI

CORACIIFORMES

TYPES OF PASSERINES

Passerines, or perching birds, are by far the most numerous order of birds, consisting of approximately 5,300 species, or half of all bird species.

Long-tailed tit

Eurasian blackbird

Bar-throated apalis

Guiana cock-of-the-rock

Bobolink

Common nightingale

Banded honeyeaters

FLIGHT STAGES OF THE EUROPEAN ROBIN

Primary coverts convert muscle power into forward movement.

Hollow wing bones attach to powerful muscles on the breast, combining lightness with strength.

Feathers are separated to reduce air resistance.

Throughout the upstroke, feathers are tucked well into the body.

The wings drop, the drag on the feathers pulling the robin forward.

Feathers twist apart to prepare for upstroke.

EVOLUTION OF GALAPAGOS FINCHES

Ancestral ground finch
Probably ate seeds

Sharp-billed finch
Pecks seabirds to drink blood

Tree finch
Eats insects

Warbler finch
Extracts insects from twigs

Woodpecker finch
Raps rotten wood for insects

Large cactus ground finch
Eats cactus flowers and seeds

Large ground finch
Cracks and eats large seeds

Tree finch
Eats plant material

Broadbills, Pittas, Ovenbirds and Tyrant Flycatchers

Location

■ Broadbills and pittas
■ Ovenbirds and tyrant flycatchers

GREEN BROADBILL
Calyptomena viridis

CASSIN'S KINGBIRD
Tyrannus vociferans

BANDED PITTA
Pitta guajana

GREAT KISKADEE
Pitangus sulphuratus

BLACK PHOEBE
Sayornis nigricans

GUIANA COCK-OF-THE-ROCK
Rupicola rupicola

EASTERN KINGBIRD
Tyrannus tyrannus

MALE WIRE-TAILED MANAKIN
Pipra filicauda

Rufous hornero

Western kingbird

Wire-tailed manakin skeleton

Lyrebirds, Scrub-birds, Larks, Wagtails and Swallows

Location

■ Larks, wagtails and swallows
■ Larks, wagtails, swallows, lyrebirds and scrub-birds

WAGTAILS
Motacilla sp.

Adult pied wagtail

Juvenile white wagtail   Adult white wagtail

WOODLARK
Lullula arborea

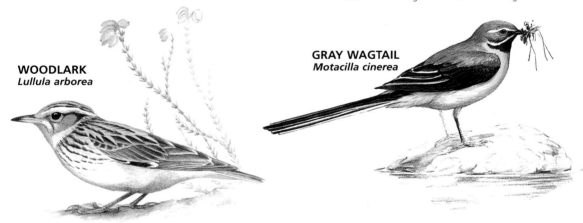

GRAY WAGTAIL
Motacilla cinerea

SKYLARKS FEEDING
Alauda arvensis

WATER PIPIT
Anthus spinoletta

Male (left) and female purple martins

Barn swallow

Male (left) and female swallows

Sand martins at nest cavity

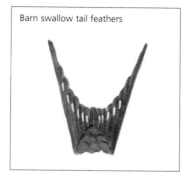
Barn swallow tail feathers

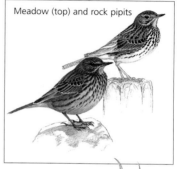
Meadow (top) and rock pipits

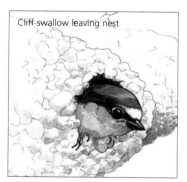
Cliff swallow leaving nest

Prairie (left) and northern horned larks

Superb lyrebird displaying tail plumage

Cliff swallow tail feathers

British yellow wagtail head

Continental yellow wagtail head

Cuckooshrikes, Bulbuls, Leafbirds, Shrikes, Vangas and Waxwings

Location

- Cuckooshrikes, bulbuls and leafbirds
- Shrikes, vangas and waxwings
- Cuckooshrikes, bulbuls, leafbirds, shrikes, vangas and waxwings

RED-VENTED BULBUL
Pycnonotus cafer

SCARLET MINIVET
Pericrocotus flammeus

YELLOW-EYED CUCKOOSHRIKE
Coracina lineata

BOHEMIAN WAXWING
Bombycilla garrulus

SCARLET MINIVET IN FLIGHT
Scarlet minivets fold their wings briefly after completing a strong downstroke, to save energy before the next upstroke.

LONG-TAILED SHRIKE
Lanius schach

GOLDEN-FRONTED LEAFBIRD
Chloropsis aurifrons

WHITE-CHEEKED BULBUL
Pycnonotus leucogenys

ORANGE-BELLIED LEAFBIRD
Chloropsis hardwickii

Mockingbirds, Accentors, Dippers, Thrushes, Babblers and Wrens

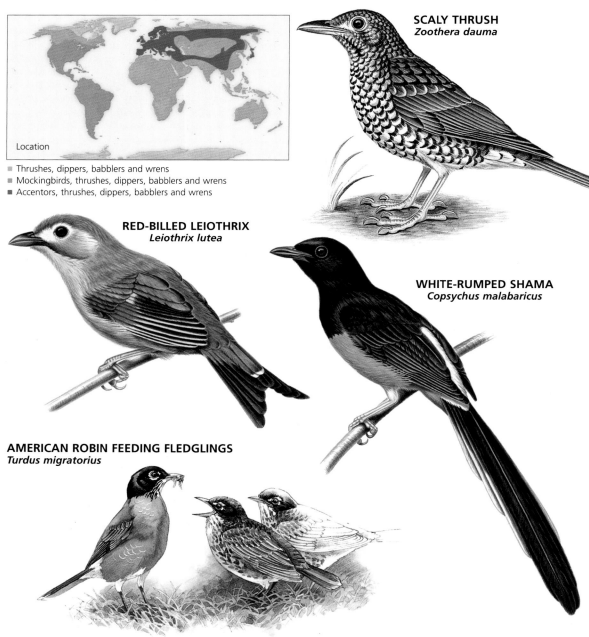

Location

- Thrushes, dippers, babblers and wrens
- Mockingbirds, thrushes, dippers, babblers and wrens
- Accentors, thrushes, dippers, babblers and wrens

SCALY THRUSH
Zoothera dauma

RED-BILLED LEIOTHRIX
Leiothrix lutea

WHITE-RUMPED SHAMA
Copsychus malabaricus

AMERICAN ROBIN FEEDING FLEDGLINGS
Turdus migratorius

Male American redstart on branch

Wren singing on branch

Wren at entrance of dome-shaped nest

Male black redstart

Canyon wren on rock

Male European robin 'redbreast'

Stonechats and whinchat (bottom)

Mockingbird in flight

Eurasian blackbird

Male western bluebird

Eastern bluebird

Dunnock

Warblers, Flycatchers, Fairywrens, Logrunners and Monarchs

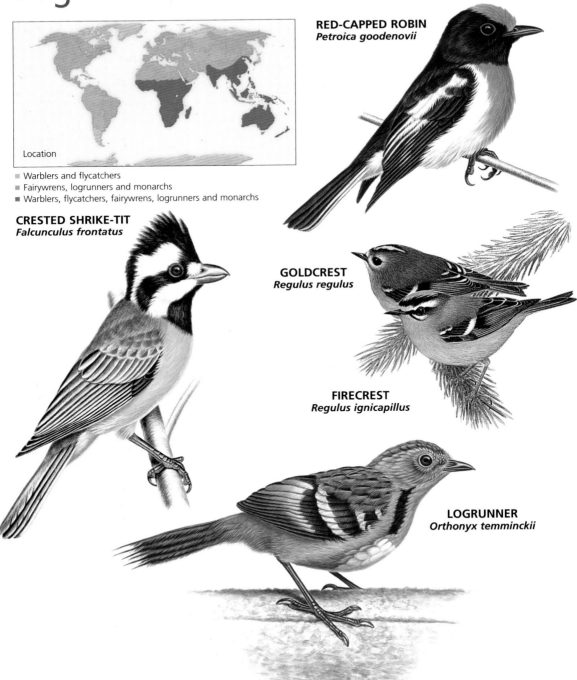

Location

- Warblers and flycatchers
- Fairywrens, logrunners and monarchs
- Warblers, flycatchers, fairywrens, logrunners and monarchs

RED-CAPPED ROBIN
Petroica goodenovii

CRESTED SHRIKE-TIT
Falcunculus frontatus

GOLDCREST
Regulus regulus

FIRECREST
Regulus ignicapillus

LOGRUNNER
Orthonyx temminckii

GARDEN WARBLER
Sylvia borin

RUFOUS-BELLIED NILTAVA
Niltava sundara

SPECTACLED MONARCH
Monarcha trivirgatus

GOLDEN-CROWNED KINGLET
Regulus satrapa

SPLENDID FAIRYWREN
Malurus splendens

325

Tits, Nuthatches, Treecreepers and Honeyeaters

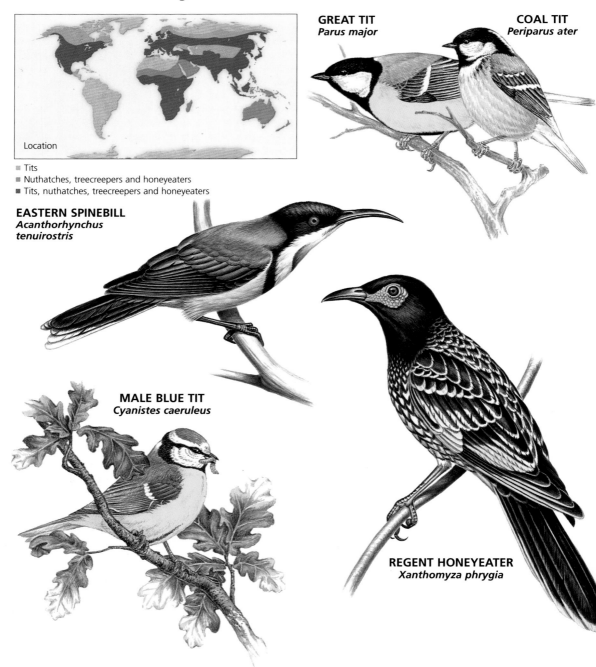

Location

- ▪ Tits
- ▪ Nuthatches, treecreepers and honeyeaters
- ▪ Tits, nuthatches, treecreepers and honeyeaters

GREAT TIT
Parus major

COAL TIT
Periparus ater

EASTERN SPINEBILL
Acanthorhynchus tenuirostris

MALE BLUE TIT
Cyanistes caeruleus

REGENT HONEYEATER
Xanthomyza phrygia

SPOTTED PARDALOTE
Pardalotus punctatus

REGAL SUNBIRD
Cinnyris regius

EURASIAN NUTHATCH
Sitta europaea

WHITE-BREASTED NUTHATCH
Sitta carolinensis

PLAIN TITMOUSE
Parus inornatus

Family group of long-tailed tits

Chickadee resting on branch

South-western (top) and eastern tufted titmice

Buntings, Tanagers, Vireos and Wood Warblers

Location

- Buntings
- Buntings, tanagers, vireos and wood warblers

FEMALE CHESTNUT-SIDED WARBLER
Dendroica pensylvanica

Breeding plumage

RED-LEGGED HONEYCREEPER
Cyanerpes cyaneus

JUVENILE WHITE-THROATED SPARROW
Zonotrichia albicollis

SEVEN-COLORED TANAGER
Tangara fastuosa

Female

Male in molt

Male with winter plumage

SCARLET TANAGERS
Piranga olivacea

PAINTED BUNTING
Passerina ciris

Female northern cardinal head

Male northern cardinal

Comparison of male pyrrhuloxia (left) and northern cardinal heads

House sparrows

Slate-colored (top) and Oregon juncos

Scarlet tanager nest with eggs

Female house sparrow dustbathing

Northern parula at nest dug into moss

Adult white-striped sparrow head

Warbling (top) and Philadelphia vireos

Virginia's warbler on branch

Nashville warbler at nest entrance

329

Finches

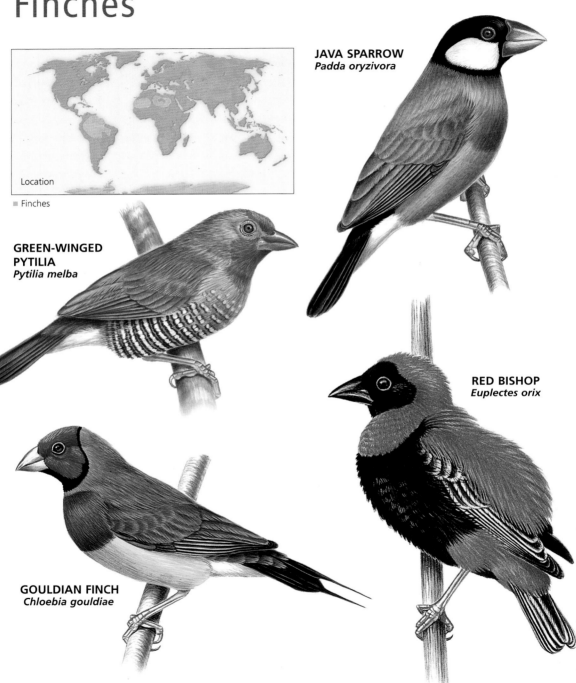

Location

■ Finches

JAVA SPARROW
Padda oryzivora

GREEN-WINGED PYTILIA
Pytilia melba

RED BISHOP
Euplectes orix

GOULDIAN FINCH
Chloebia gouldiae

Eurasian bullfinch at birth
Naked skin; eyes closed

At six days
Sparse covering of down; eyes open

At 28 days
Feathers grow out of tracts along wings.

At nine months
Bullfinch fully grown

EUROPEAN GOLDFINCH
Carduelis carduelis

Male   Female

EUROPEAN GREENFINCHES
Carduelis chloris

CHAFFINCH
Fringilla coelebs

Starlings, Magpie-larks and Icterids

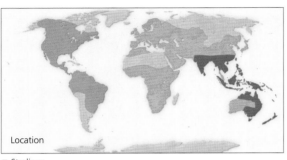

Location

- Starlings
- Magpie-larks (Old World) and icterids (New World)
- Starlings, magpie-larks and icterids

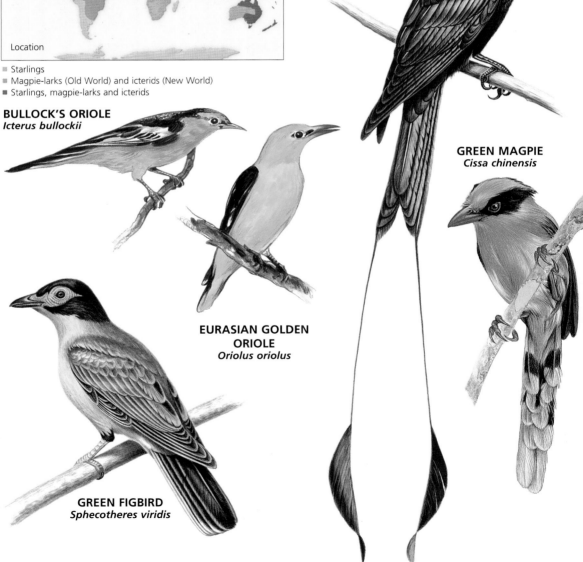

GREATER RACKET-TAILED DRONGO
Dicrurus paradiseus

BULLOCK'S ORIOLE
Icterus bullockii

GREEN MAGPIE
Cissa chinensis

EURASIAN GOLDEN ORIOLE
Oriolus oriolus

GREEN FIGBIRD
Sphecotheres viridis

Passerines

BIRDS

SUPERB STARLING
Lamprotornis superbus

EUROPEAN STARLINGS
Sturnus vulgaris

Summer plumage

Winter plumage

ORANGE ORIOLE
Icterus auratus

YELLOW ORIOLE
Icterus nigrogularis

Common grackle egg

Baltimore oriole feeding young

Tricolored (top) and red-winged blackbirds

Bowerbirds and Birds-of-Paradise

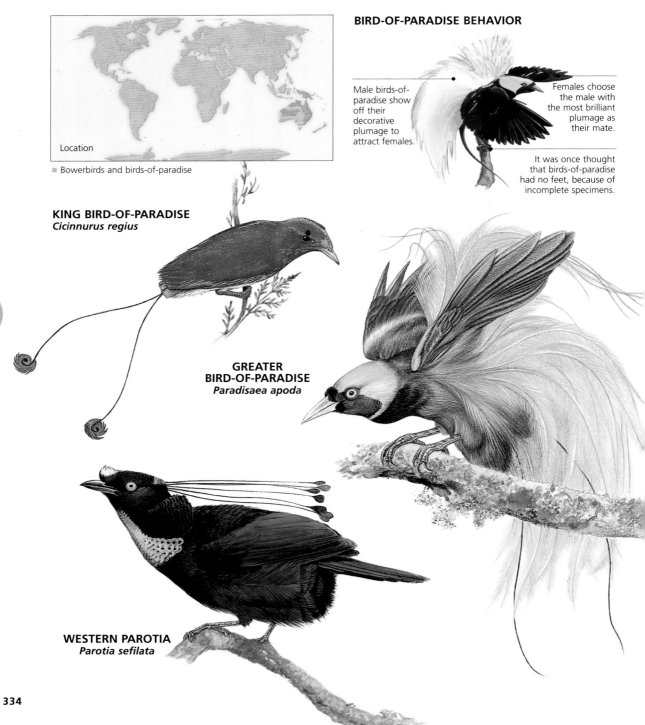

Location

■ Bowerbirds and birds-of-paradise

BIRD-OF-PARADISE BEHAVIOR

Male birds-of-paradise show off their decorative plumage to attract females.

Females choose the male with the most brilliant plumage as their mate.

It was once thought that birds-of-paradise had no feet, because of incomplete specimens.

KING BIRD-OF-PARADISE
Cicinnurus regius

GREATER BIRD-OF-PARADISE
Paradisaea apoda

WESTERN PAROTIA
Parotia sefilata

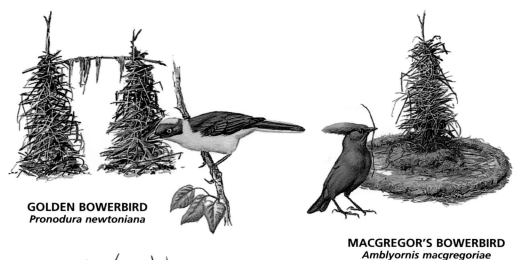

GOLDEN BOWERBIRD
Pronodura newtoniana

MACGREGOR'S BOWERBIRD
Amblyornis macgregoriae

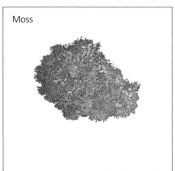

SATIN BOWERBIRD
*Ptilonorhynchus
violaceus*

TOOTH-BILLED BOWERBIRD
Ailuroedus dentirostris

BOWER DECORATIONS

Male bowerbirds create elaborate bowers, or courtship display areas, in one of four different types—court, mat, maypole or avenue. These are decorated with a variety of found objects, from feathers to clothes-pegs.

Moss

Twigs

Feather

Crows and Jays

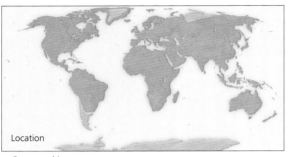

Location

■ Crows and jays

RED-BILLED CHOUGH
Pyrrhocorax pyrrhocorax

AMERICAN CROW FLIGHT
Corvus brachyrhynchos

Crows beat their wings
evenly, with strong
downstrokes, to maintain
a level flight path.

YELLOW-BILLED CHOUGH
Pyrrhocorax graculus

ROOKS
Corvus frugilegus

Adult

Juvenile

BLUE JAY
Cyanocitta cristata

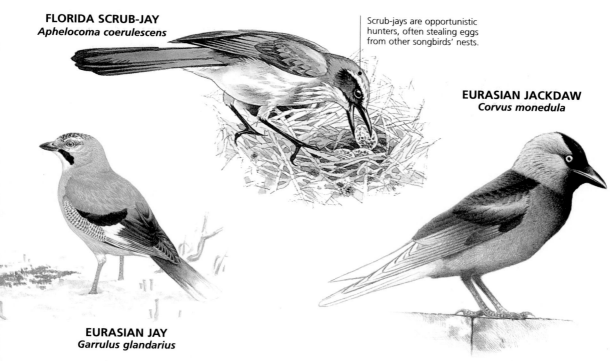

FLORIDA SCRUB-JAY
Aphelocoma coerulescens

Scrub-jays are opportunistic hunters, often stealing eggs from other songbirds' nests.

EURASIAN JACKDAW
Corvus monedula

EURASIAN JAY
Garrulus glandarius

Yellow-billed magpie head

Blue jay head showing markings

American crow (top) and common raven

Male pinyon jay on pine branch

Crested jay

Common raven head

337

Classifying Reptiles

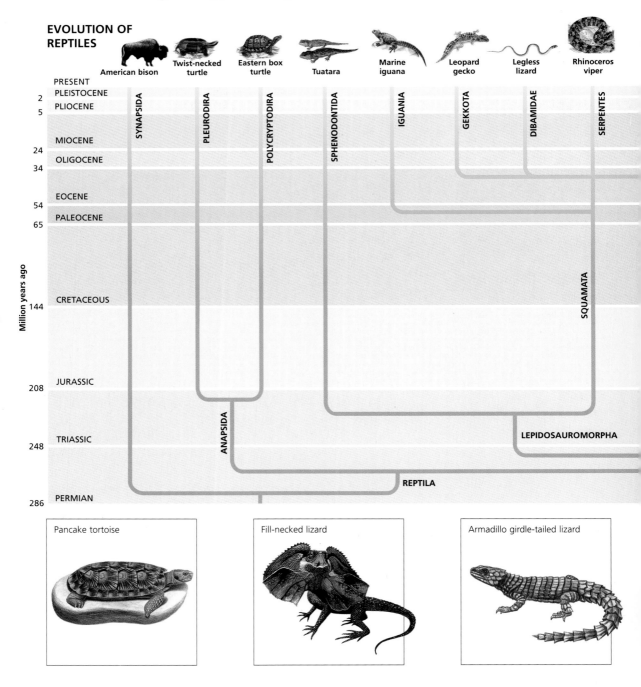

EVOLUTION OF REPTILES

American bison | Twist-necked turtle | Eastern box turtle | Tuatara | Marine iguana | Leopard gecko | Legless lizard | Rhinoceros viper

Million years ago

PRESENT
PLEISTOCENE
2
PLIOCENE
5

MIOCENE
24
OLIGOCENE
34

EOCENE
54
PALEOCENE
65

CRETACEOUS
144

JURASSIC
208

TRIASSIC
248

PERMIAN
286

SYNAPSIDA
PLEURODIRA
POLYCRYPTODIRA
SPHENODONTIDA
IGUANIA
GEKKOTA
DIBAMIDAE
SERPENTES
SQUAMATA
ANAPSIDA
LEPIDOSAUROMORPHA
REPTILA

Pancake tortoise

Fill-necked lizard

Armadillo girdle-tailed lizard

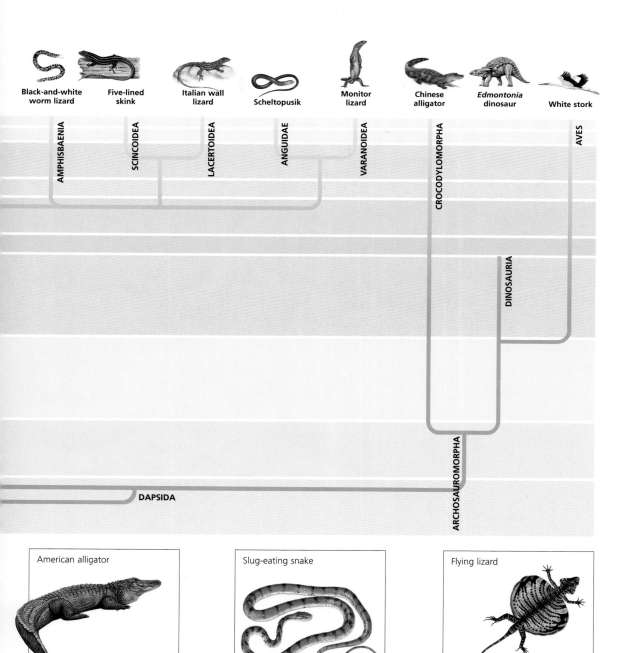

Black-and-white worm lizard

Five-lined skink

Italian wall lizard

Scheltopusik

Monitor lizard

Chinese alligator

Edmontonia dinosaur

White stork

AMPHISBAENIA

SCINCOIDEA

LACERTOIDEA

ANGUIDAE

VARANOIDEA

CROCODYLOMORPHA

AVES

DINOSAURIA

ARCHOSAUROMORPHA

DAPSIDA

American alligator

Slug-eating snake

Flying lizard

Reptile Characteristics

TYPES OF REPTILES

REPTILE SKIN

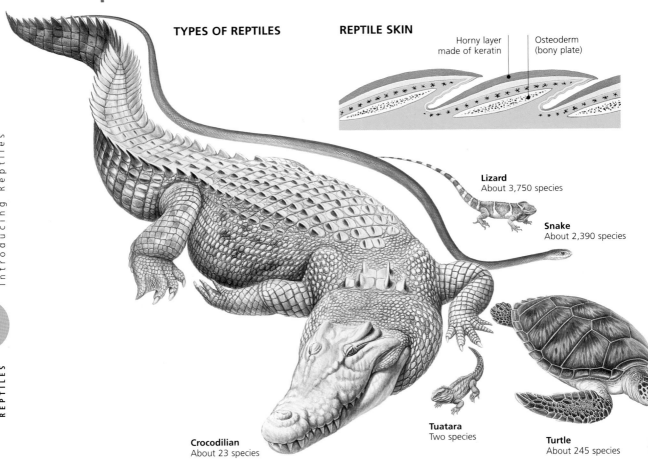

Horny layer made of keratin

Osteoderm (bony plate)

Lizard
About 3,750 species

Snake
About 2,390 species

Tuatara
Two species

Turtle
About 245 species

Crocodilian
About 23 species

Reptile Egg

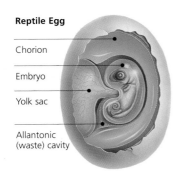

Chorion

Embryo

Yolk sac

Allantonic (waste) cavity

REPTILE SCALES

Granular scales

Keeled scales

Smooth scales

REGULATING TEMPERATURE
Sand lizard

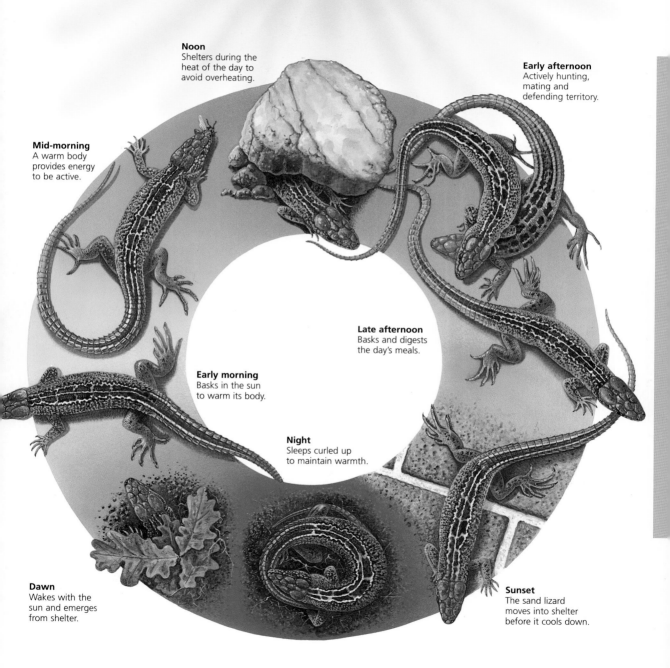

Noon
Shelters during the heat of the day to avoid overheating.

Early afternoon
Actively hunting, mating and defending territory.

Mid-morning
A warm body provides energy to be active.

Late afternoon
Basks and digests the day's meals.

Early morning
Basks in the sun to warm its body.

Night
Sleeps curled up to maintain warmth.

Dawn
Wakes with the sun and emerges from shelter.

Sunset
The sand lizard moves into shelter before it cools down.

Evolution of Reptiles

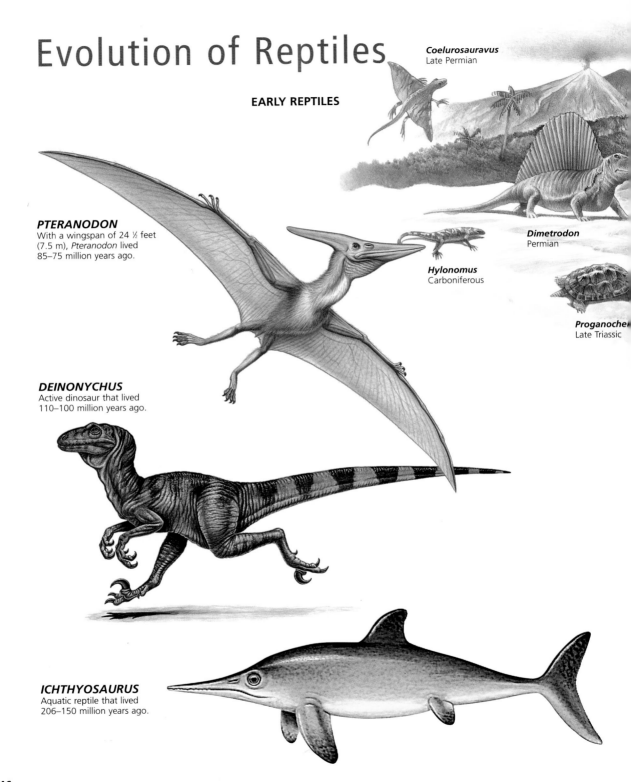

Coelurosauravus
Late Permian

EARLY REPTILES

PTERANODON
With a wingspan of 24 ½ feet
(7.5 m), *Pteranodon* lived
85–75 million years ago.

Hylonomus
Carboniferous

Dimetrodon
Permian

Proganoche
Late Triassic

DEINONYCHUS
Active dinosaur that lived
110–100 million years ago.

ICHTHYOSAURUS
Aquatic reptile that lived
206–150 million years ago.

Pteranodon
Late Cretaceous

Stegosaurus
Late Jurassic

Pachyrachis
Early Cretaceous

Planocephalosaurus
Late Triassic

Deinosuchus
Cretaceous

Archelon
Late Cretaceous

Ichthyosaurus
Early Jurassic

Elasmosaurus
Late Cretaceous

Turtles and Tortoises

LOGGERHEAD TURTLE 'CRYING'

Sea turtles, such as loggerhead turtles, 'weep' from glands around their eyes to remove salt from them. These tears are washed away by the ocean, so are visible only on land.

TORTOISE BURROW

Gopher tortoises dig burrows in the desert to escape the intensity of the heat of the day and chill of the night.

TURTLE REPRODUCTION

Mating
While some turtle species mate at sea, others mate on land. Almost all turtle species, however, lay their eggs on land.

Laying eggs
Most turtles and tortoises lay their eggs in a nest chamber, which keeps the developing eggs at a constant temperature.

Hatching out
Newly hatched turtles dig themselves out of the nest chamber. They are independent from the moment they are born.

Turtles and Tortoises

REPTILES

TUCKED AWAY

Most turtles are able to draw their necks and limbs completely into their shell, protecting themselves from predators.

FROM ABOVE AND BELOW

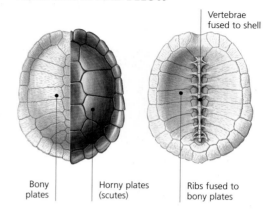

Vertebrae fused to shell

Bony plates

Horny plates (scutes)

Ribs fused to bony plates

TURTLE SHELLS

Land tortoise
Domed shell

Semi-terrestrial turtle
Flattened shell

Pond turtle
Small flattened shell

Sea turtle
Streamlined shell

TURTLE LIMBS

Land tortoise

Pond turtle

Sea turtle

Side-necked Turtles and Land Tortoises

Location

- Side-necked turtles
- Land tortoises
- Side-necked turtles and land tortoises

BIG-HEADED TURTLE
Platysternon megacephalum

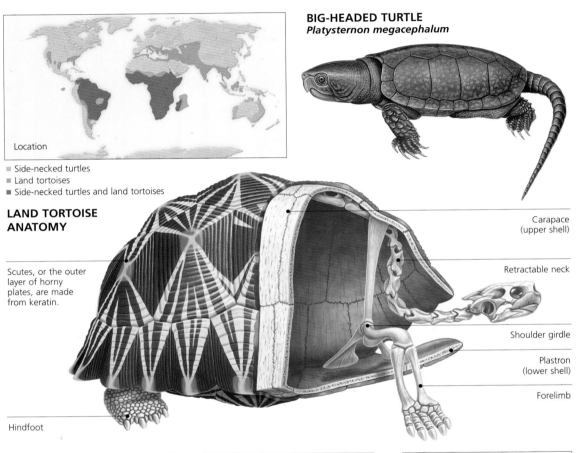

LAND TORTOISE ANATOMY

Scutes, or the outer layer of horny plates, are made from keratin.

Hindfoot

Carapace (upper shell)

Retractable neck

Shoulder girdle

Plastron (lower shell)

Forelimb

Smallest (Madagascan spider tortoise) to largest (Galápagos tortoise)

Radiated tortoise

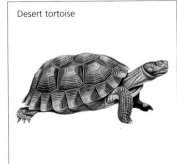

Desert tortoise

YELLOW-SPOTTED AMAZON RIVER TURTLE
Podocnemis unifilis

TWIST-NECKED TURTLE
Platemys platycephala platycephala

SADDLEBACK TORTOISES
Geochelone sp.

Cactus plants form a major part of the saddleback tortoise's diet when there is no groundwater to drink.

The tortoises can stretch their necks out to reach high-growing food, or tuck them fully into their shells.

By lowering their hindlegs to the ground, saddleback tortoises can move their inflexible bodies higher.

The domed shells of many large tortoise species are honeycombed, not solid, to be less heavy.

Turtles

Location

■ Turtles

MALE PAINTED TURTLE
Chrysemys picta belli

MALAYAN SNAIL-EATING TURTLE
Malayemys subtrijuga

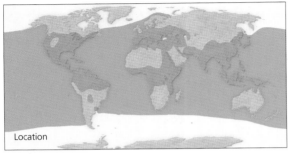

**SOUTHERN LOGGERHEAD
MUSK TURTLE**
Sternotherus minor minor

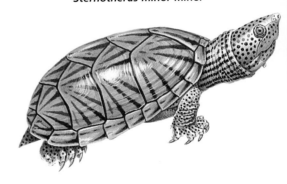

EASTERN BOX TURTLE
Terrapene carolina carolina

SPINED TURTLE
Heosemys spinosa

Raised nostrils on snout allow turtle to breathe out of water.

Forelimbs have evolved into flippers.

Marine turtles build up large fat reserves to feed off when migrating to breed and lay eggs.

PACIFIC HAWKSBILL TURTLE
Eretmochelys imbricata bissa

EASTERN SPINY SOFTSHELL TURTLE
Apalone spinifera spinifera

SOFTSHELL TURTLES

Unlike most turtle species, softshell turtles do not have hard carapaces, but fattened shells of leathery skin. Well-adapted for life in the water, they have webbed feet and a snorkle-shaped snout.

Crocodilians

CROCODILE HATCHING

GUARDING THE NEST
Female crocodilians cover their egg nests with warm rotting plant material, and do not leave until their young are hatched.

CARRYING YOUNG
Alligator hatchlings are carried in the mother's mouth to a nearby pond. The female will protect the hatchlings from predators for several weeks.

CROCODILIAN SKULLS

Temporal opening (fenestra)

External nostril

Eye socket

Temporal opening (fenestra)

External nostril

Eye socket

Lower jaw opening

CROCODILE GAITS

Crawling

Walking

Galloping

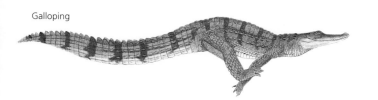

ANKLE COMPARISON

Crocodilian	Typical reptile

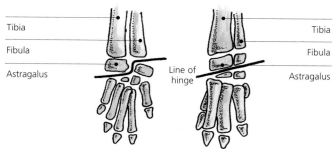

Tibia

Fibula

Astragalus

Line of hinge

Tibia

Fibula

Astragalus

SNOUT COMPARISON

American crocodile

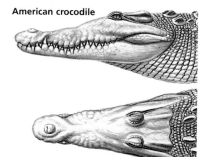

Black caiman

Gharial

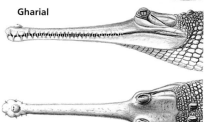

BREATHING HALF-SUBMERGED

External nostrils remain above water surface.

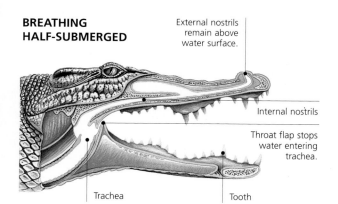

Internal nostrils

Throat flap stops water entering trachea.

Trachea

Tooth

CROCODILE TOOTH

Gum

Successional tooth

Tooth socket

Dentine

Functional tooth

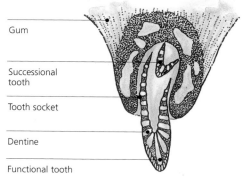

Alligators and Caimans

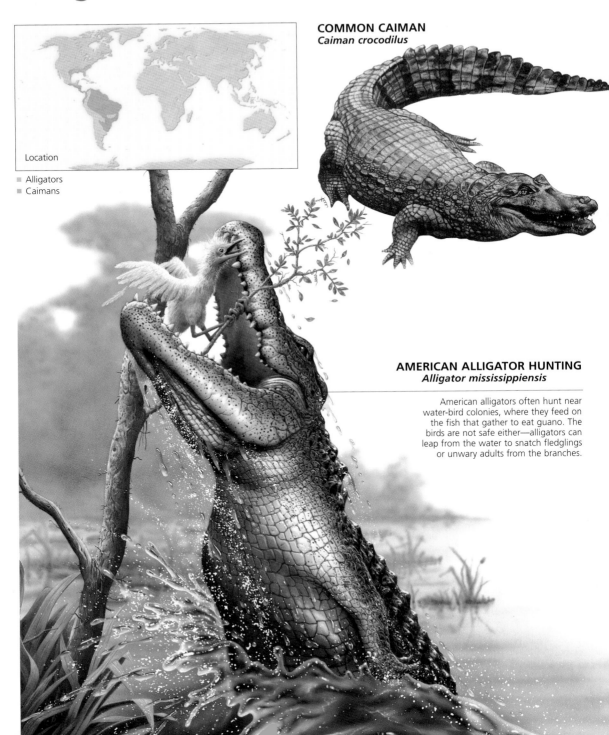

Location
- Alligators
- Caimans

COMMON CAIMAN
Caiman crocodilus

AMERICAN ALLIGATOR HUNTING
Alligator mississippiensis

American alligators often hunt near water-bird colonies, where they feed on the fish that gather to eat guano. The birds are not safe either—alligators can leap from the water to snatch fledglings or unwary adults from the branches.

BLACK CAIMAN
Melanosuchus niger

CHINESE ALLIGATOR
Alligator sinensis

SCHNEIDER'S DWARF CAIMAN
Paleosuchus trigonatus

CUVIER'S DWARF CAIMAN
Paleosuchus palpebrosus

Crocodiles and Gharials

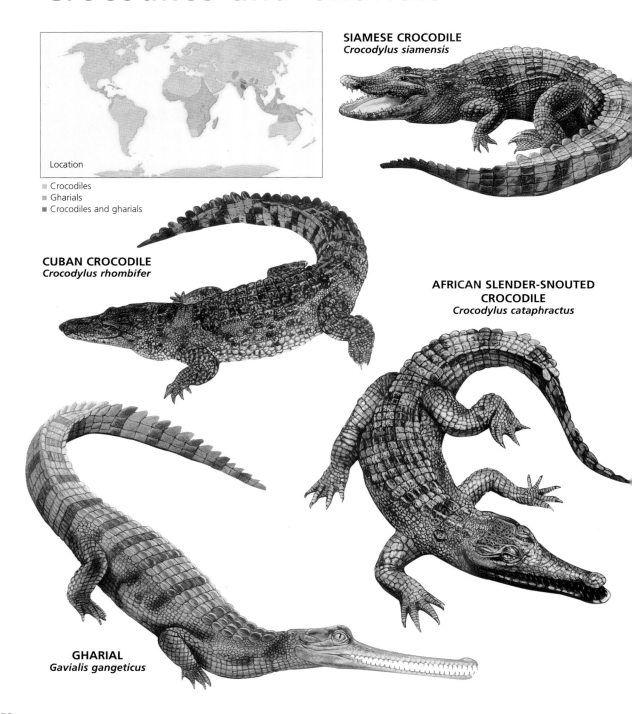

Location

- Crocodiles
- Gharials
- Crocodiles and gharials

SIAMESE CROCODILE
Crocodylus siamensis

CUBAN CROCODILE
Crocodylus rhombifer

AFRICAN SLENDER-SNOUTED CROCODILE
Crocodylus cataphractus

GHARIAL
Gavialis gangeticus

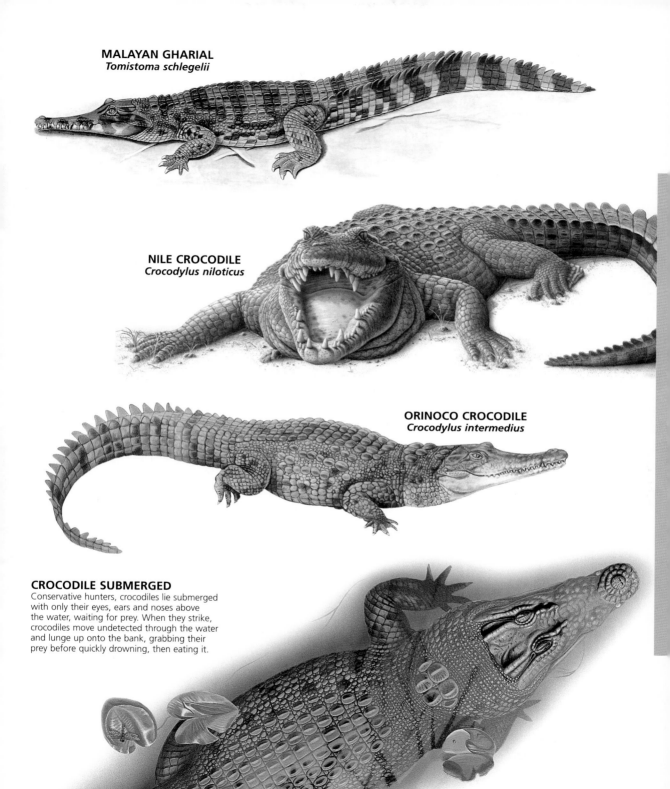

MALAYAN GHARIAL
Tomistoma schlegelii

NILE CROCODILE
Crocodylus niloticus

ORINOCO CROCODILE
Crocodylus intermedius

CROCODILE SUBMERGED

Conservative hunters, crocodiles lie submerged with only their eyes, ears and noses above the water, waiting for prey. When they strike, crocodiles move undetected through the water and lunge up onto the bank, grabbing their prey before quickly drowning, then eating it.

Amphisbaenians and Tuatara

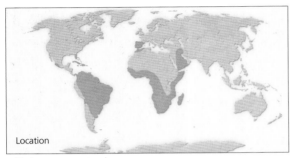

Location

■ Tuatara
■ Amphisbaenians (worm-lizards)

ADAPTATIONS FOR DIGGING

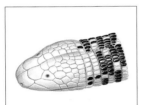

Round-snouted
Push forward, then around

Spade-snouted
Push forward, then up

Chisel-snouted
Rotate in opposing directions

Keel-snouted
Push forward, then sideways

BLACK AND WHITE WORM-LIZARD
Amphisbaena fuliginosa

KEEL-SNOUTED WORM-LIZARD
Family *Amphisbaenidae*

SPADE-SNOUTED WORM-LIZARD
Family *Amphisbaenidae*

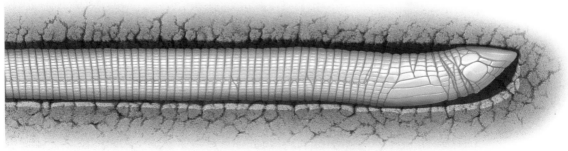

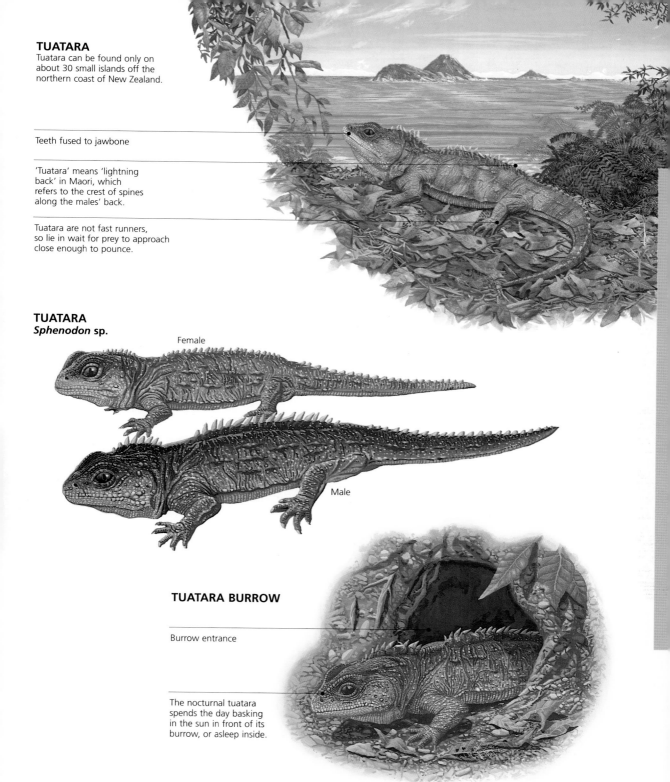

TUATARA

Tuatara can be found only on about 30 small islands off the northern coast of New Zealand.

Teeth fused to jawbone

'Tuatara' means 'lightning back' in Maori, which refers to the crest of spines along the males' back.

Tuatara are not fast runners, so lie in wait for prey to approach close enough to pounce.

TUATARA
Sphenodon sp.

Female

Male

TUATARA BURROW

Burrow entrance

The nocturnal tuatara spends the day basking in the sun in front of its burrow, or asleep inside.

Lizards

EVOLUTION OF LIZARDS

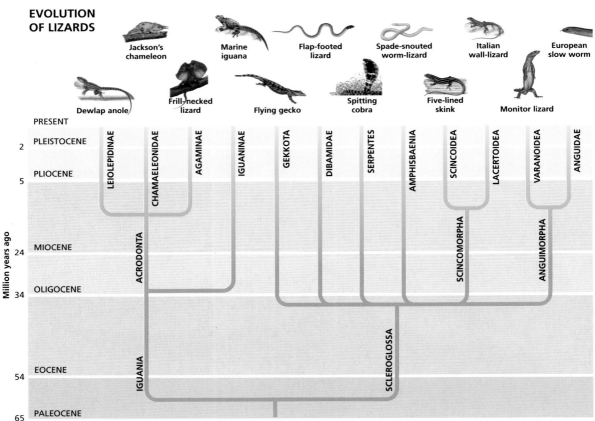

Jackson's chameleon

Marine iguana

Flap-footed lizard

Spade-snouted worm-lizard

Italian wall-lizard

European slow worm

Dewlap anole

Frill-necked lizard

Flying gecko

Spitting cobra

Five-lined skink

Monitor lizard

Million years ago

PRESENT
PLEISTOCENE
2
PLIOCENE
5

MIOCENE
24

OLIGOCENE
34

EOCENE
54

PALEOCENE
65

LEIOLEPIDINAE
CHAMAELEONIDAE
AGAMINAE
IGUANINAE
GEKKOTA
DIBAMIDAE
SERPENTES
AMPHISBAENIA
SCINCOIDEA
LACERTOIDEA
VARANOIDEA
ANGUIDAE

ACRODONTA
SCINCOMORPHA
ANGUIMORPHA

IGUANIA
SCLEROGLOSSA

LIZARD TOES

Lizard toes have adapted to specific uses and environments, such as burrowing or climbing.

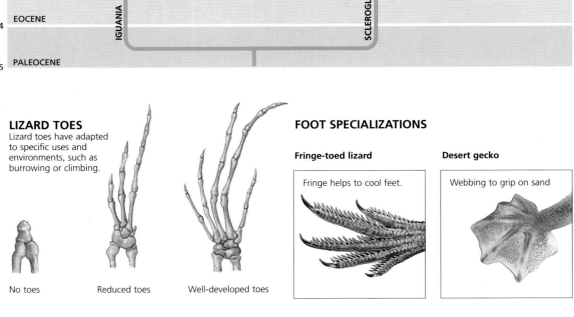

No toes

Reduced toes

Well-developed toes

FOOT SPECIALIZATIONS

Fringe-toed lizard

Fringe helps to cool feet.

Desert gecko

Webbing to grip on sand

362

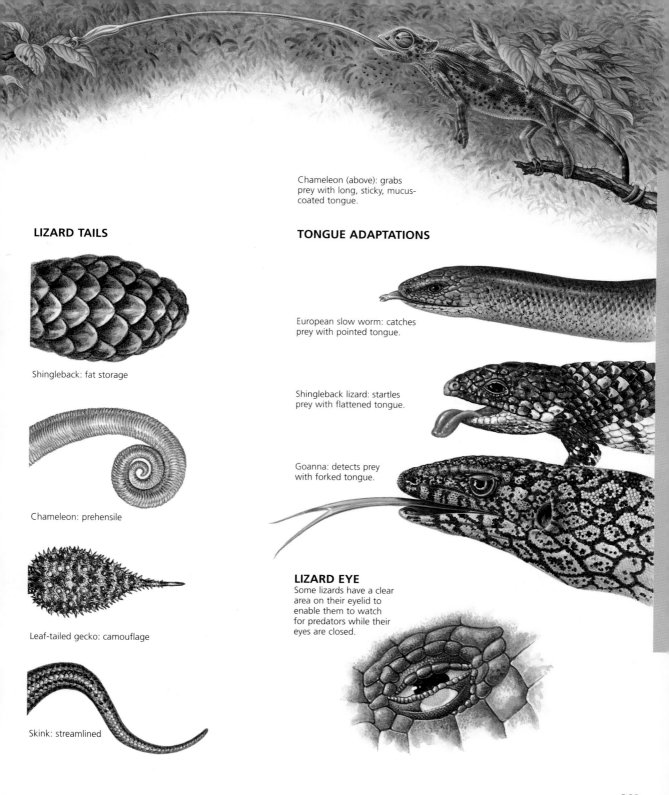

Chameleon (above): grabs prey with long, sticky, mucus-coated tongue.

LIZARD TAILS

Shingleback: fat storage

Chameleon: prehensile

Leaf-tailed gecko: camouflage

Skink: streamlined

TONGUE ADAPTATIONS

European slow worm: catches prey with pointed tongue.

Shingleback lizard: startles prey with flattened tongue.

Goanna: detects prey with forked tongue.

LIZARD EYE
Some lizards have a clear area on their eyelid to enable them to watch for predators while their eyes are closed.

Iguanids

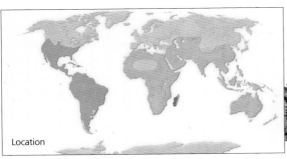

Location

- Agamids and chameleons
- Iguanas and anoles
- Agamids, chameleons, iguanas and anoles

HELMETED IGUANA HEAD

The function of the casque, or crest, extending from head to neck, is unknown.

Like chameleons, helmeted iguanas can change color.

Pouch under the chin expands when iguana defends its territory.

COLLARED LIZARD BEHAVIOR

Crouches down

Pushes up to appear bigger

Crouches down again

Rhinoceros iguanas change color with the heat of the sun.

Thorny devil

Basilisk running on water

Regal horned lizard squirting blood

Common iguana head

Bearded dragon expanding throat

MARINE IGUANA
Amblyrhynchus cristatus

CUBAN BROWN ANOLE
Anolis sagrei sagrei

JACKSON'S CHAMELEON
Chamaeleo jacksonii

Agamids and Chameleons

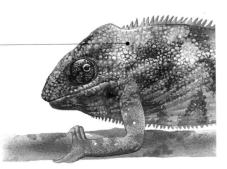

CHANGING COLOR

Male chameleons flush with color to warn rival males off their territories. This species changes from camouflaging green to an angry red.

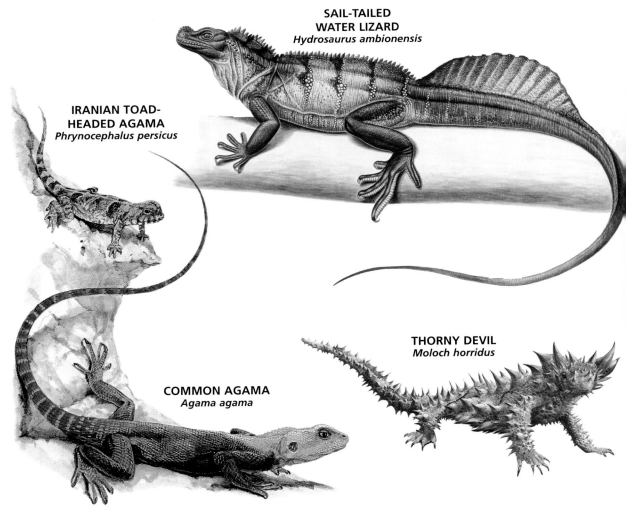

SAIL-TAILED WATER LIZARD
Hydrosaurus ambionensis

IRANIAN TOAD-HEADED AGAMA
Phrynocephalus persicus

THORNY DEVIL
Moloch horridus

COMMON AGAMA
Agama agama

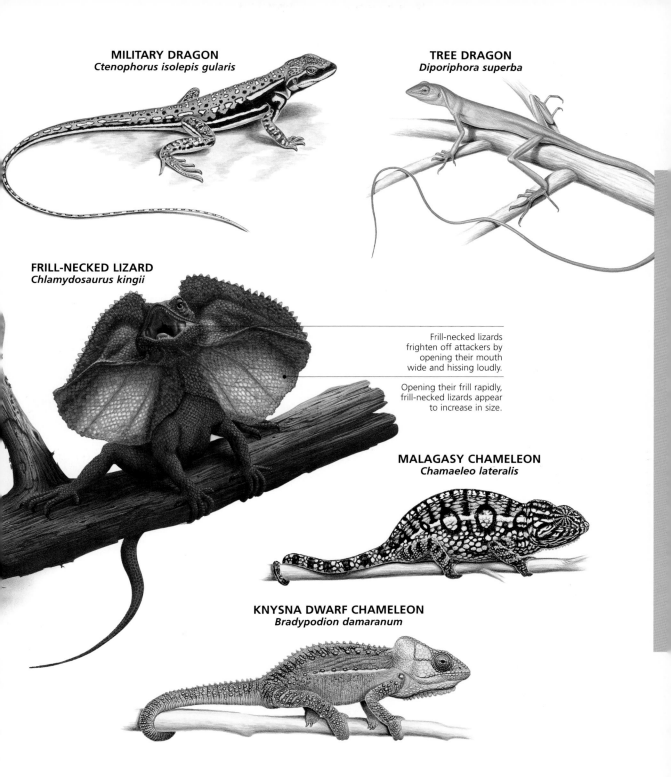

MILITARY DRAGON
Ctenophorus isolepis gularis

TREE DRAGON
Diporiphora superba

FRILL-NECKED LIZARD
Chlamydosaurus kingii

Frill-necked lizards frighten off attackers by opening their mouth wide and hissing loudly.

Opening their frill rapidly, frill-necked lizards appear to increase in size.

MALAGASY CHAMELEON
Chamaeleo lateralis

KNYSNA DWARF CHAMELEON
Bradypodion damaranum

Iguanas and Anoles

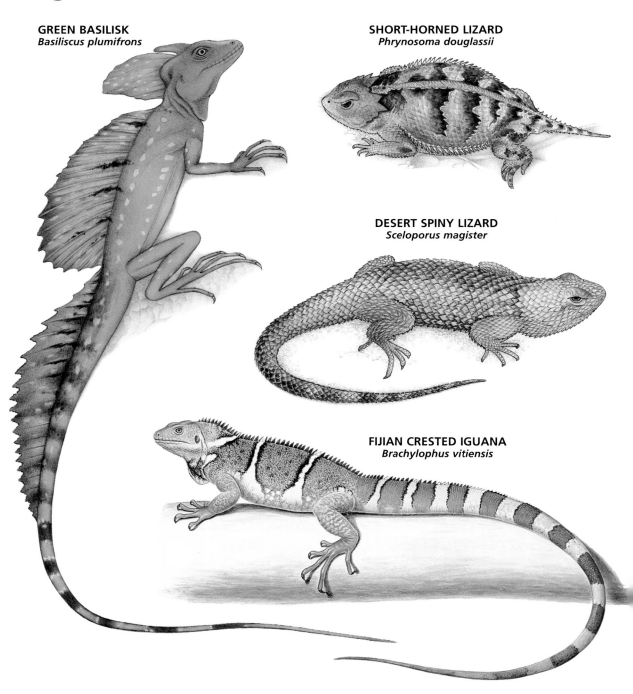

GREEN BASILISK
Basiliscus plumifrons

SHORT-HORNED LIZARD
Phrynosoma douglassii

DESERT SPINY LIZARD
Sceloporus magister

FIJIAN CRESTED IGUANA
Brachylophus vitiensis

COLLARED LIZARD
Crotaphytus collaris

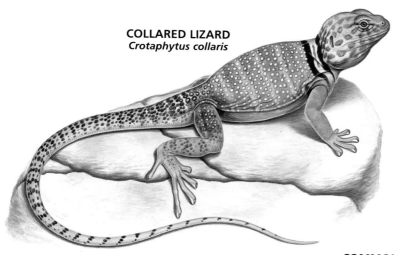

COMMON BASILISK
Basiliscus basiliscus

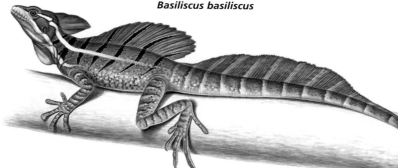

MARINE IGUANAS

Marine iguanas of the Galápagos Islands are the only species of lizard that depend on the sea to survive.

Male marine iguanas change color to green and red to show their readiness to mate.

The cold ocean waters mean that the iguanas spend hours basking in the sun to regain enough warmth to feed again.

Geckos

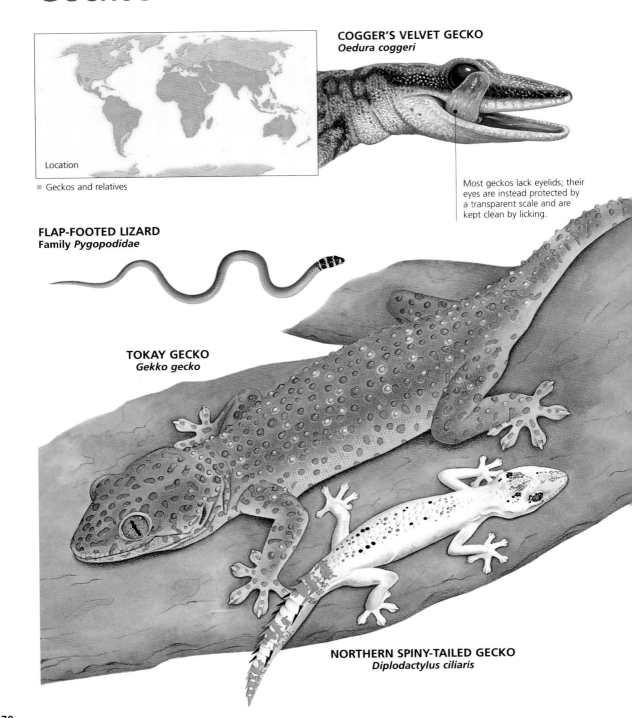

Location
■ Geckos and relatives

COGGER'S VELVET GECKO
Oedura coggeri

Most geckos lack eyelids; their eyes are instead protected by a transparent scale and are kept clean by licking.

FLAP-FOOTED LIZARD
Family *Pygopodidae*

TOKAY GECKO
Gekko gecko

NORTHERN SPINY-TAILED GECKO
Diplodactylus ciliaris

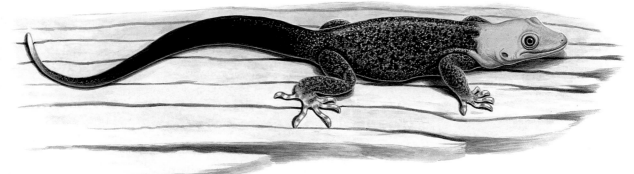

YELLOW-HEADED GECKO
Gonatodes albogularis fuscus

LEOPARD GECKO
Eublepharus macularius

GRAY'S BOW-FINGERED GECKO
Cyrtodactylus pulchellus

COMMON WONDER GECKO
Teratoscincus scincus

Geckos

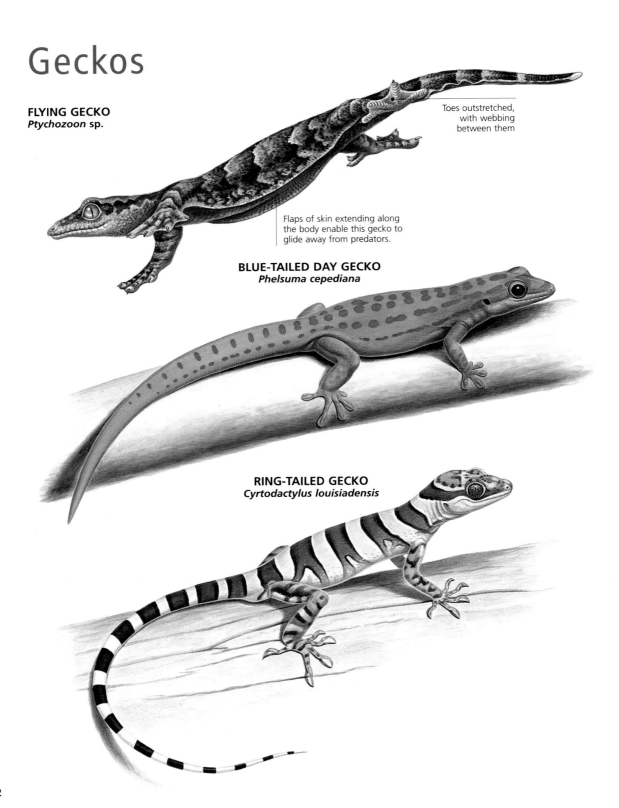

FLYING GECKO
Ptychozoon sp.

Toes outstretched,
with webbing
between them

Flaps of skin extending along
the body enable this gecko to
glide away from predators.

BLUE-TAILED DAY GECKO
Phelsuma cepediana

RING-TAILED GECKO
Cyrtodactylus louisiadensis

SOUTHERN SPOTTED VELVET GECKO
Oedura tryoni

MOURNING GECKO
Lepidodactylus lugubris

Female mourning geckos are able to reproduce without fertilizing their eggs, and therefore without needing a male.

COMMON WALL GECKO
Tarentola mauritanica

The common wall gecko, or Moorish gecko, is one of only four species still found in Europe.

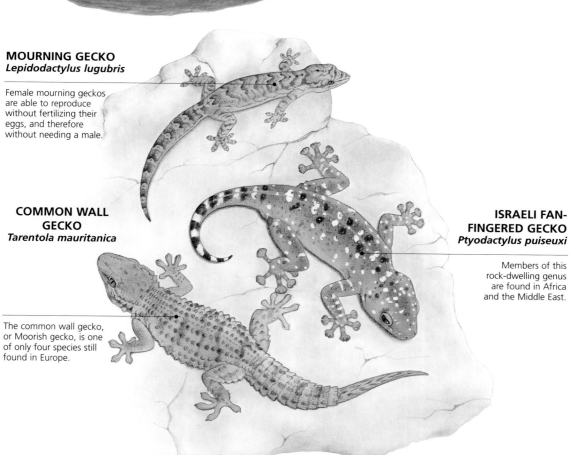

ISRAELI FAN-FINGERED GECKO
Ptyodactylus puiseuxi

Members of this rock-dwelling genus are found in Africa and the Middle East.

Beaded Lizards and Monitors

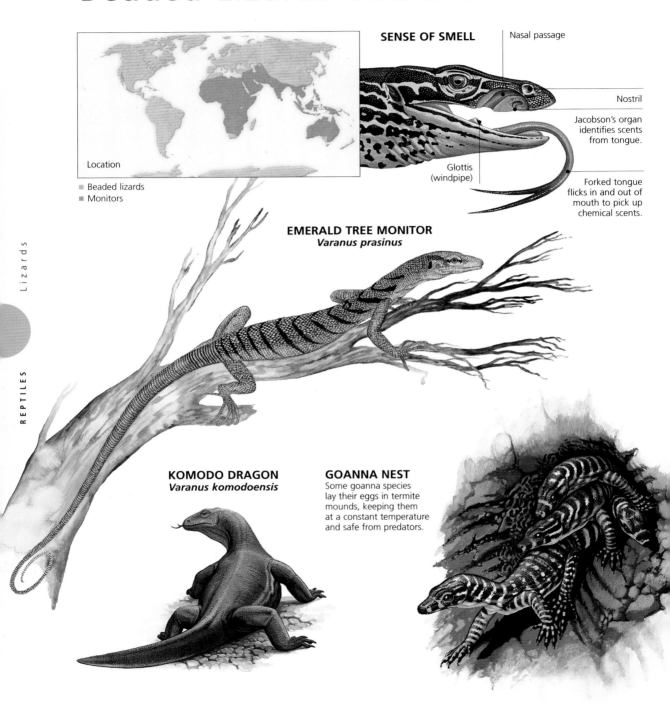

SENSE OF SMELL

Nasal passage

Nostril

Jacobson's organ identifies scents from tongue.

Glottis (windpipe)

Forked tongue flicks in and out of mouth to pick up chemical scents.

Location

■ Beaded lizards
■ Monitors

EMERALD TREE MONITOR
Varanus prasinus

KOMODO DRAGON
Varanus komodoensis

GOANNA NEST
Some goanna species lay their eggs in termite mounds, keeping them at a constant temperature and safe from predators.

CROCODILE MONITOR
Varanus salvadorii

PERENTIE
Varanus giganteus

GOULD'S MONITOR
Varanus gouldii

GILA MONSTER
Heloderma suspectum

Skinklike Lizards

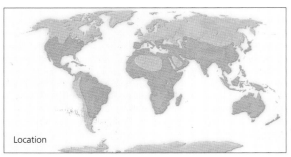

Location

■ Skinklike lizards

BLUE-TONGUE SKINK WITH YOUNG
Tiliqua scincoides

VIVIPAROUS LIZARD
Lacerta vivipara

SCHELTOPUSIK
Ophisaurus apodus

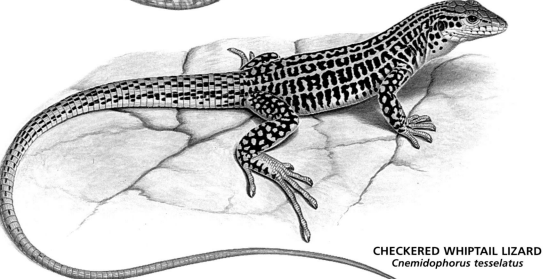

CHECKERED WHIPTAIL LIZARD
Cnemidophorus tesselatus

GRANITE NIGHT LIZARD
Xantusia henshawi

PINK-TONGUED SKINK
Cyclodomorphus gerrardii

**SOUTH AMERICAN SKINKLIKE
LIZARD**
Kentropix calcarata

RAINBOW LIZARD
Cnemidophorus lemniscatus

Skinklike Lizards

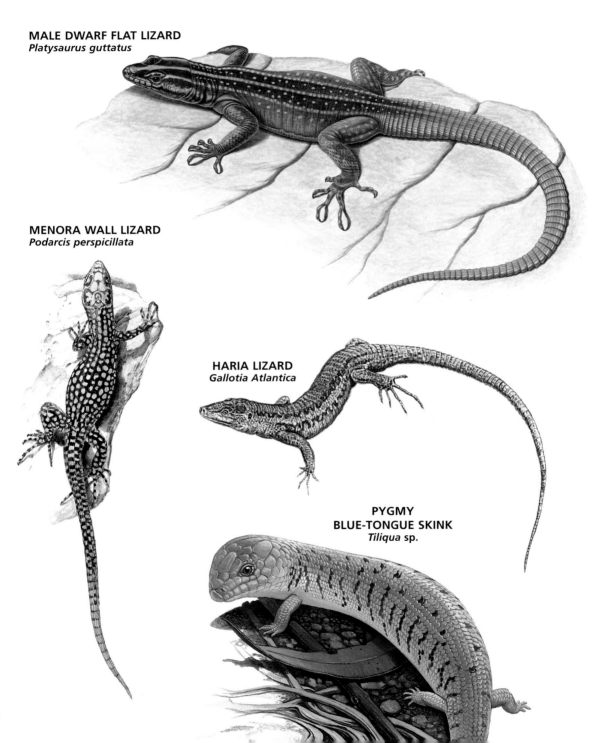

MALE DWARF FLAT LIZARD
Platysaurus guttatus

MENORA WALL LIZARD
Podarcis perspicillata

HARIA LIZARD
Gallotia Atlantica

**PYGMY
BLUE-TONGUE SKINK**
Tiliqua sp.

JUNGLE RUNNER
Ameiva ameiva

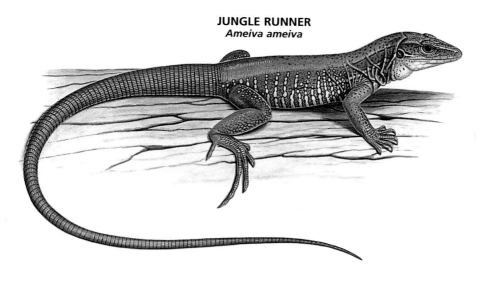

JUVENILE FIVE-LINED SKINK
Eumeces fasciatus

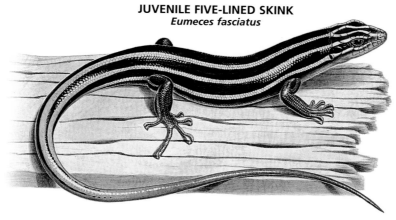

ITALIAN WALL LIZARD
Podarcis sicula

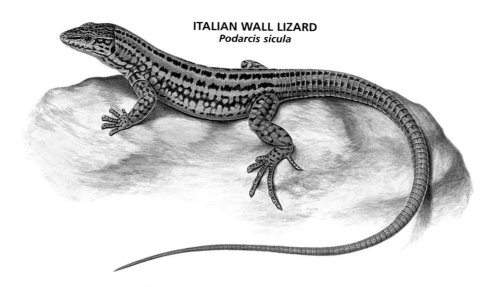

Snakes

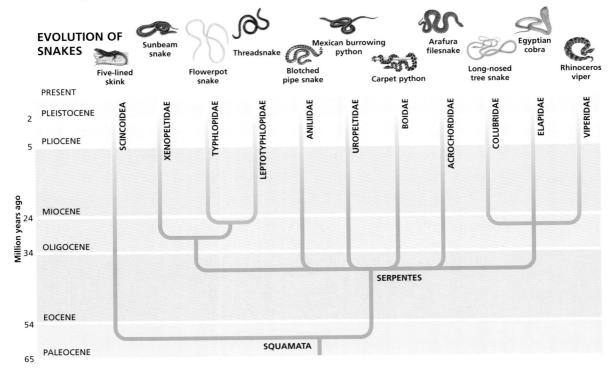

EVOLUTION OF SNAKES

Five-lined skink · Sunbeam snake · Flowerpot snake · Threadsnake · Blotched pipe snake · Mexican burrowing python · Carpet python · Arafura filesnake · Long-nosed tree snake · Egyptian cobra · Rhinoceros viper

SCINCOIDEA · XENOPELTIDAE · TYPHLOPIDAE · LEPTOTYPHLOPIDAE · ANILIIDAE · UROPELTIDAE · BOIDAE · ACROCHORDIDAE · COLUBRIDAE · ELAPIDAE · VIPERIDAE

PRESENT
PLEISTOCENE
2
PLIOCENE
5

Million years ago

MIOCENE
24
OLIGOCENE
34

SERPENTES

EOCENE
54
PALEOCENE
65

SQUAMATA

SNAKE ANATOMY

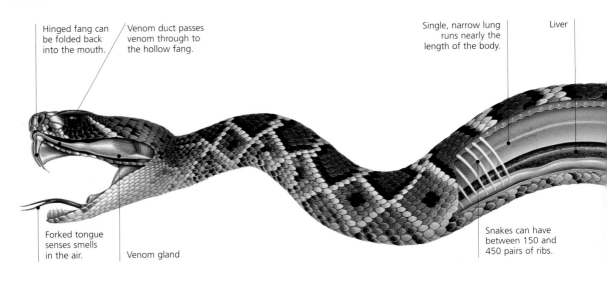

Hinged fang can be folded back into the mouth.

Venom duct passes venom through to the hollow fang.

Single, narrow lung runs nearly the length of the body.

Liver

Forked tongue senses smells in the air.

Venom gland

Snakes can have between 150 and 450 pairs of ribs.

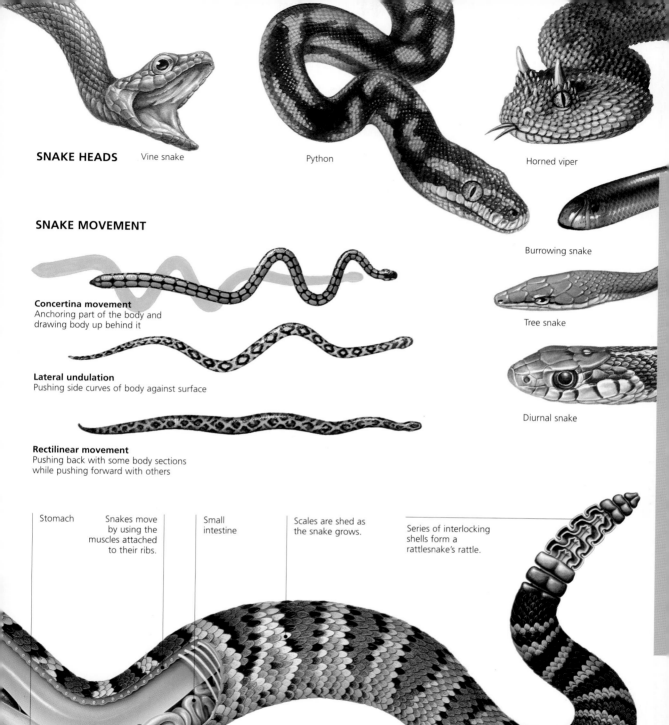

SNAKE HEADS

Vine snake

Python

Horned viper

Burrowing snake

Tree snake

Diurnal snake

SNAKE MOVEMENT

Concertina movement
Anchoring part of the body and
drawing body up behind it

Lateral undulation
Pushing side curves of body against surface

Rectilinear movement
Pushing back with some body sections
while pushing forward with others

Stomach

Snakes move
by using the
muscles attached
to their ribs.

Small
intestine

Scales are shed as
the snake grows.

Series of interlocking
shells form a
rattlesnake's rattle.

Wormsnakes and Pipe Snakes

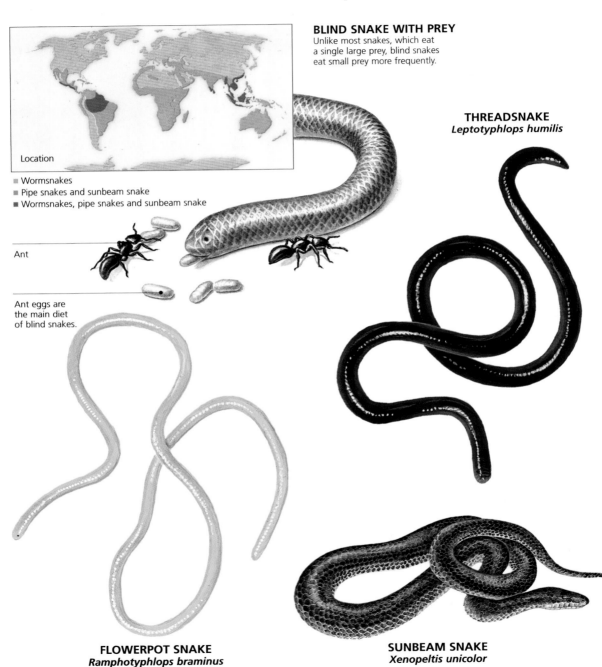

BLIND SNAKE WITH PREY
Unlike most snakes, which eat a single large prey, blind snakes eat small prey more frequently.

Location

- Wormsnakes
- Pipe snakes and sunbeam snake
- Wormsnakes, pipe snakes and sunbeam snake

Ant

Ant eggs are the main diet of blind snakes.

THREADSNAKE
Leptotyphlops humilis

FLOWERPOT SNAKE
Ramphotyphlops braminus

SUNBEAM SNAKE
Xenopeltis unicolor

MEXICAN BURROWING PYTHON
Loxocemus bicolor

**SOUTH AMERICAN CORAL
PIPE SNAKE**
Anilius scytale

**DRUMMOND HAY'S
EARTH SNAKE**
Rhinophis drummondhayi

BLOTCHED PIPE SNAKE
Cylindrophus maculatus

RED CYLINDER SNAKE
Cylindrophis rufus

Boas and Pythons

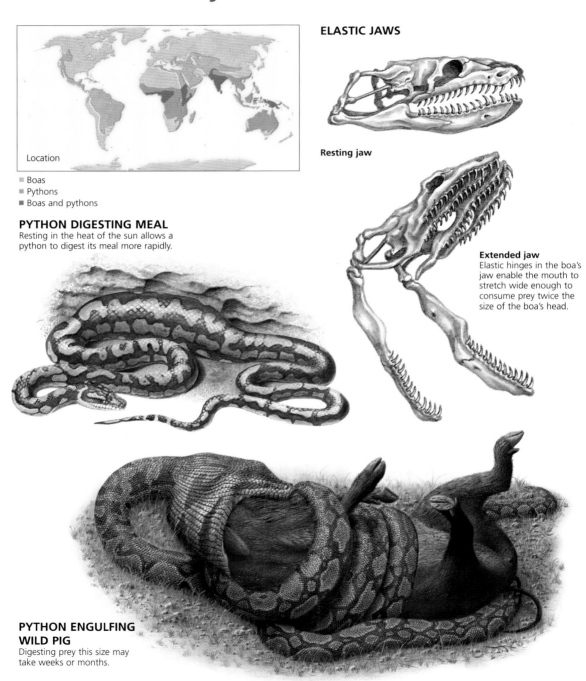

Location

- Boas
- Pythons
- Boas and pythons

ELASTIC JAWS

Resting jaw

PYTHON DIGESTING MEAL
Resting in the heat of the sun allows a
python to digest its meal more rapidly.

Extended jaw
Elastic hinges in the boa's
jaw enable the mouth to
stretch wide enough to
consume prey twice the
size of the boa's head.

**PYTHON ENGULFING
WILD PIG**
Digesting prey this size may
take weeks or months.

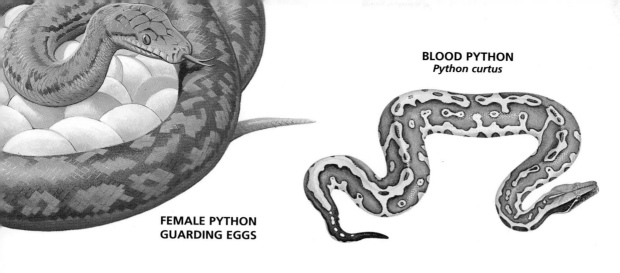

FEMALE PYTHON GUARDING EGGS

BLOOD PYTHON
Python curtus

SIZE COMPARISON

Anaconda (species of boa): 33 feet (10 m)

Boa constrictor: 14½ feet (4.5 m)

Eastern diamondback rattlesnake: 7 feet (2.2 m)

Yellow-bellied sea snake: 2½ feet (0.8 m)

BRAZILIAN RAINBOW BOA
Epicrates cenchria cenchria

CARPET PYTHON
Morelia spilota

Colubrid Snakes

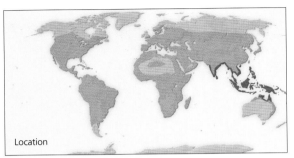

**WHITE-BELLIED
MANGROVE SNAKE**
Fordonia leucobalia

ARAFURA FILESNAKE
Acrochordus arafurae

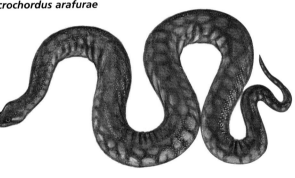

LONG-NOSED TREE SNAKE
Ahaetulla nasuta

AFRICAN TWIG SNAKE
Thelotornis capensis

LITTLE FILESNAKE
Acrochordus granulatus

COMMON GRASS SNAKE
Natrix natrix

SPOTTED HOUSE SNAKE
Lamprophis guttatus

COMMON KING SNAKE
Lampropeltis getulus

TOAD-EATER SNAKE
Xenodon rabdocephalus

Colubrid Snakes

**BLUNT-HEADED
TREE SNAKE**
Imantodes cenchoa

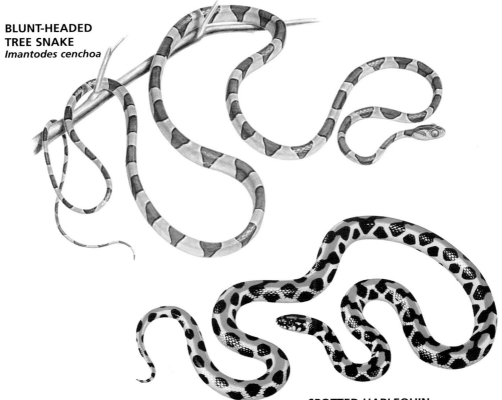

**SPOTTED HARLEQUIN
SNAKE**
Homoroselaps lacteus

MANDARIN RATSNAKE
Elaphe mandarina

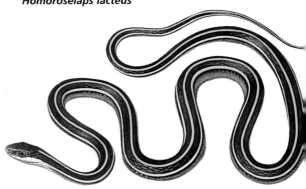

EASTERN RIBBON SNAKE
Thamnophis sauritus sauritus

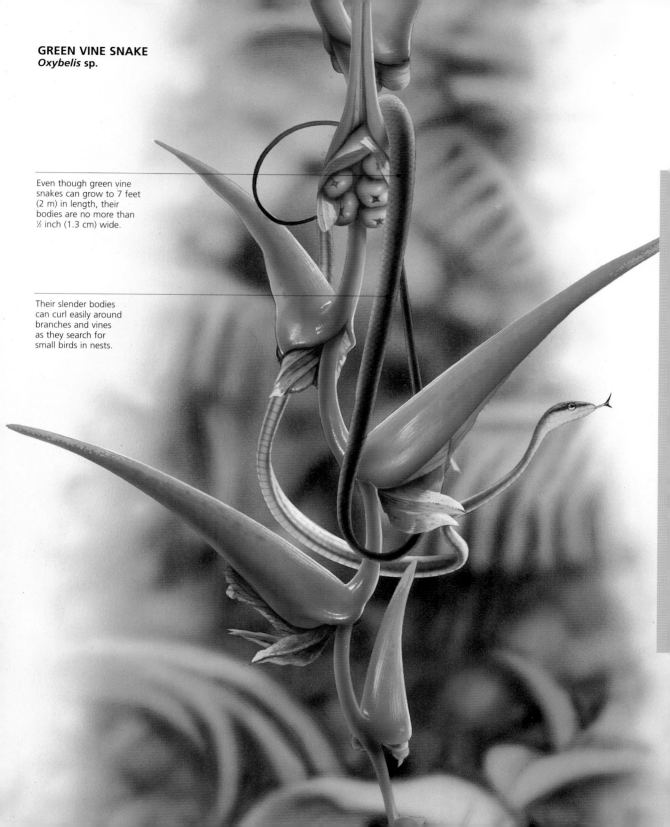

GREEN VINE SNAKE
Oxybelis sp.

Even though green vine snakes can grow to 7 feet (2 m) in length, their bodies are no more than ½ inch (1.3 cm) wide.

Their slender bodies can curl easily around branches and vines as they search for small birds in nests.

Elapid Snakes and Vipers

TYPES OF FANGS

Colubrid snakes
Rear, fixed, grooved fangs

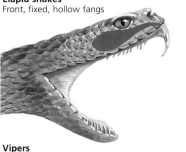

Elapid snakes
Front, fixed, hollow fangs

Vipers
Front, swinging, hollow fangs

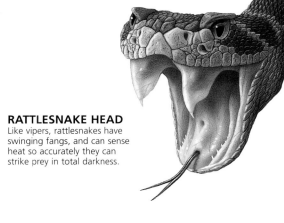

RATTLESNAKE HEAD
Like vipers, rattlesnakes have swinging fangs, and can sense heat so accurately they can strike prey in total darkness.

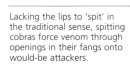

SPITTING COBRA

Lacking the lips to 'spit' in the traditional sense, spitting cobras force venom through openings in their fangs onto would-be attackers.

The cobra's hood flap is made from skin only, which extends as the elongated ribs expand when it feels angry or threatened.

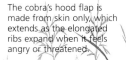

COBRA SKELETON

Elongated ribs

SIDEWINDER LOCOMOTION

The sidewinder moves across hot, loose sand by anchoring its head and tail, lifting its trunk sideways across the ground. The head and tail then join the trunk.

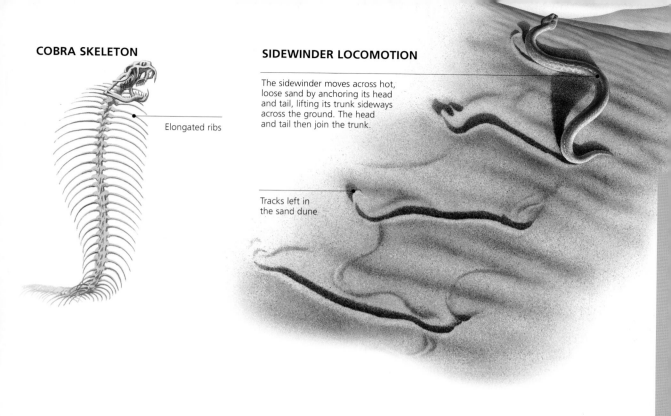

Tracks left in the sand dune

RATTLESNAKE RATTLE

Body scales

Interlocking shells

The rattle is used to distract prey, mesmerizing would-be attackers until the snake is ready to strike.

Elapid Snakes

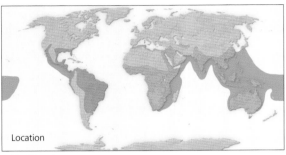

Location

■ Elapid snakes
■ Coral snakes and sea snakes

BLUE CORAL SNAKE HEAD
Maticora bivirgata

The venom duct of the blue coral snake extends almost a third of the way along the snake's body.

EGYPTIAN COBRA
Naja haje

YELLOW-LIPPED SEA KRAIT
Laticauda colubrina

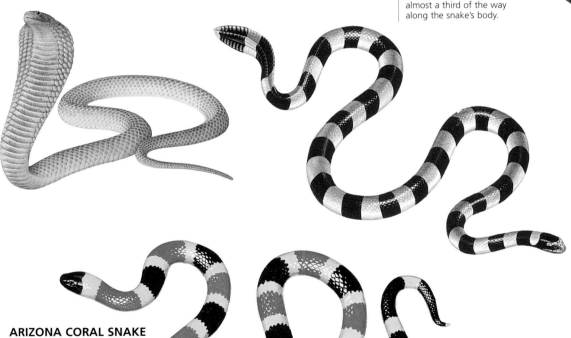

ARIZONA CORAL SNAKE
Micruroides euryxanthus euryxanthus

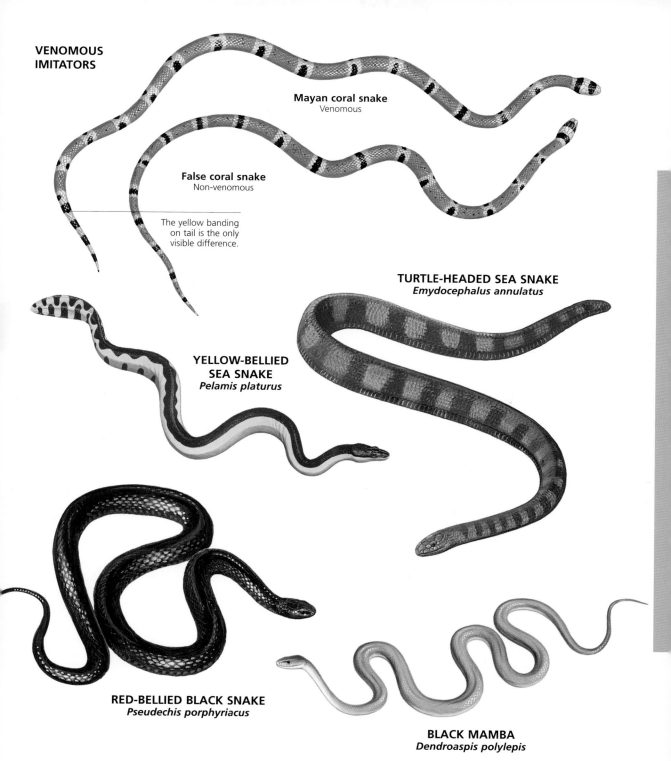

VENOMOUS IMITATORS

Mayan coral snake
Venomous

False coral snake
Non-venomous

The yellow banding on tail is the only visible difference.

TURTLE-HEADED SEA SNAKE
Emydocephalus annulatus

YELLOW-BELLIED SEA SNAKE
Pelamis platurus

RED-BELLIED BLACK SNAKE
Pseudechis porphyriacus

BLACK MAMBA
Dendroaspis polylepis

Vipers

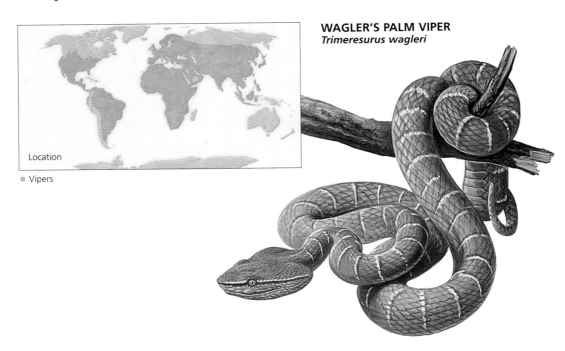

Location

- Vipers

WAGLER'S PALM VIPER
Trimeresurus wagleri

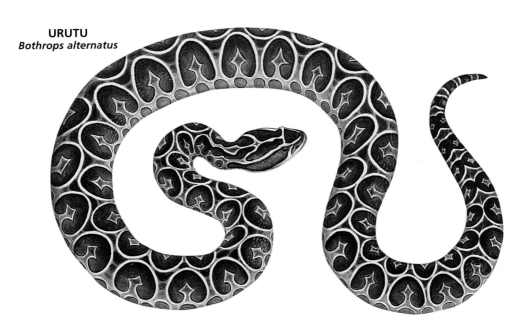

URUTU
Bothrops alternatus

RHINOCEROS VIPER
Bitis nasicornis

LANCE-HEADED RATTLESNAKE
Crotalus polystictus

COMMON DEATH ADDER
Acanthophis antarcticus

INTRODUCING AMPHIBIANS

SALAMANDERS AND NEWTS

CAECILIANS

FROGS AND TOADS

Classifying Amphibians

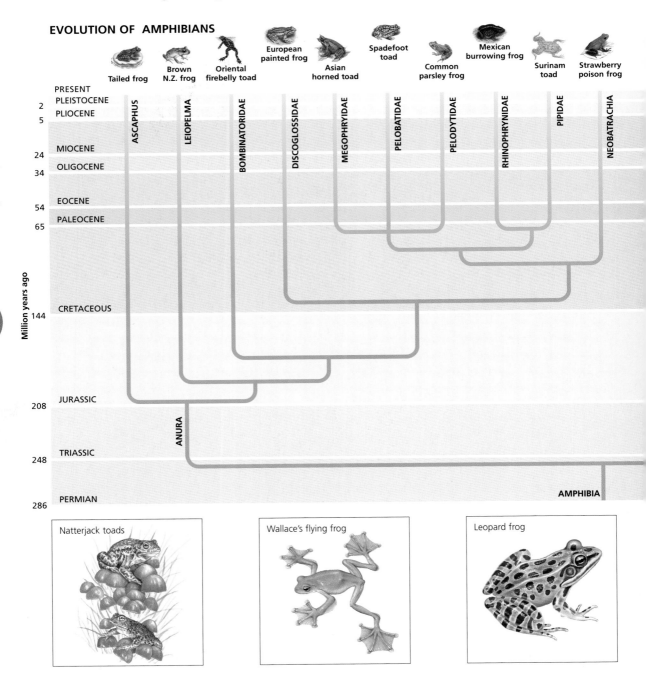

EVOLUTION OF AMPHIBIANS

Tailed frog

Brown N.Z. frog

Oriental firebelly toad

European painted frog

Asian horned toad

Spadefoot toad

Common parsley frog

Mexican burrowing frog

Surinam toad

Strawberry poison frog

Introducing Amphibians

AMPHIBIANS

Million years ago

PRESENT
PLEISTOCENE
2
PLIOCENE
5
MIOCENE
24
OLIGOCENE
34
EOCENE
54
PALEOCENE
65
CRETACEOUS
144
JURASSIC
208
TRIASSIC
248
PERMIAN
286

ASCAPHUS

LEIOPELMA

BOMBINATORIDAE

DISCOGLOSSIDAE

MEGOPHRYIDAE

PELOBATIDAE

PELODYTIDAE

RHINOPHRYNIDAE

PIPIDAE

NEOBATRACHIA

ANURA

AMPHIBIA

Natterjack toads

Wallace's flying frog

Leopard frog

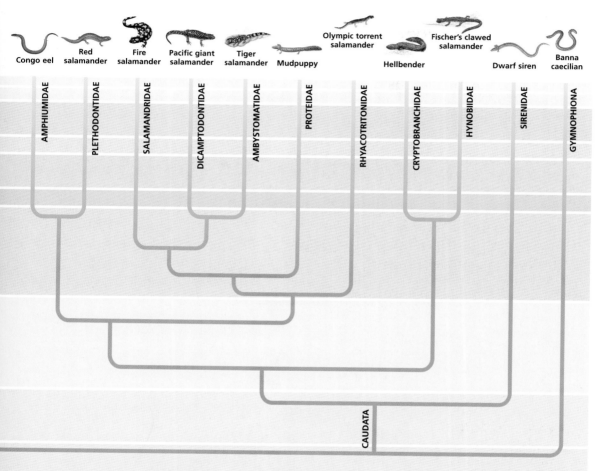

Congo eel — Red salamander — Fire salamander — Pacific giant salamander — Tiger salamander — Mudpuppy — Olympic torrent salamander — Hellbender — Fischer's clawed salamander — Dwarf siren — Banna caecilian

AMPHIUMIDAE — PLETHODONTIDAE — SALAMANDRIDAE — DICAMPTODONTIDAE — AMBYSTOMATIDAE — PROTEIDAE — RHYACOTRITONIDAE — CRYPTOBRANCHIDAE — HYNOBIIDAE — SIRENIDAE — GYMNOPHIONA

CAUDATA

Due to the incomplete fossil record and changing classification theories for amphibians, the dates indicated by the branches of the graphs are at times approximations.

Poison frog in a bromeliad

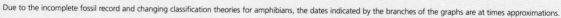

Mexican axolotyl

Terrestrial caecilian

Salamanders and Newts

EVOLUTION OF SALAMANDERS AND NEWTS

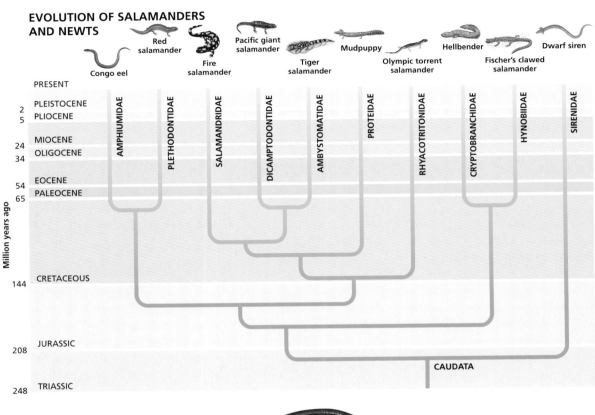

Congo eel
Red salamander
Fire salamander
Pacific giant salamander
Tiger salamander
Mudpuppy
Olympic torrent salamander
Hellbender
Fischer's clawed salamander
Dwarf siren

PRESENT
PLEISTOCENE
2
5 PLIOCENE
MIOCENE
24
34 OLIGOCENE
54 EOCENE
65 PALEOCENE
Million years ago
144 CRETACEOUS
208 JURASSIC
248 TRIASSIC

AMPHIUMIDAE
PLETHODONTIDAE
SALAMANDRIDAE
DICAMPTODONTIDAE
AMBYSTOMATIDAE
PROTEIDAE
RHYACOTRITONIDAE
CRYPTOBRANCHIDAE
HYNOBIIDAE
SIRENIDAE

CAUDATA

SALAMANDER LIFECYCLE

While lifecycles vary greatly across salamander species, the basic lifecycle, with its aquatic and terrestrial stages, is shown.

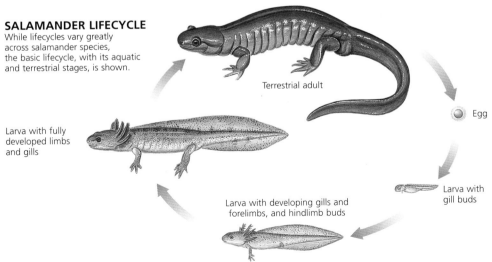

Terrestrial adult

Egg

Larva with fully developed limbs and gills

Larva with developing gills and forelimbs, and hindlimb buds

Larva with gill buds

FIRE SALAMANDER

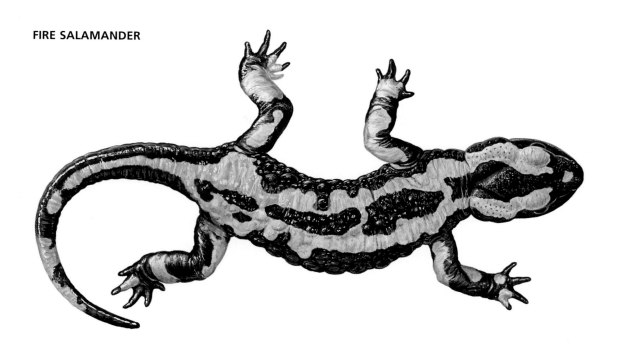

FIRE SALAMANDER
SKELETON

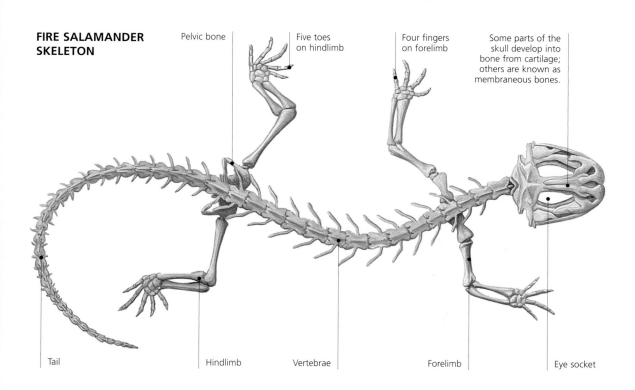

Pelvic bone

Five toes
on hindlimb

Four fingers
on forelimb

Some parts of the
skull develop into
bone from cartilage;
others are known as
membraneous bones.

Tail

Hindlimb

Vertebrae

Forelimb

Eye socket

Salamanders and Newts

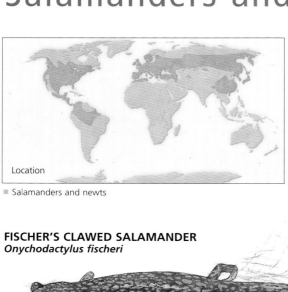

Location

■ Salamanders and newts

FIRE SALAMANDER
Salamandra salamandra

FISCHER'S CLAWED SALAMANDER
Onychodactylus fischeri

BLUE-SPOTTED SALAMANDER
Ambystoma laterale

PACIFIC GIANT SALAMANDER
Dicamptodon ensatus

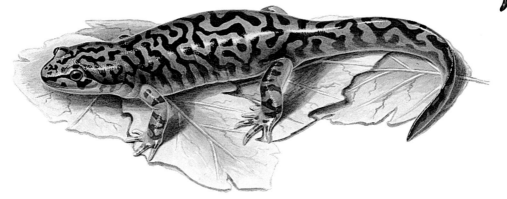

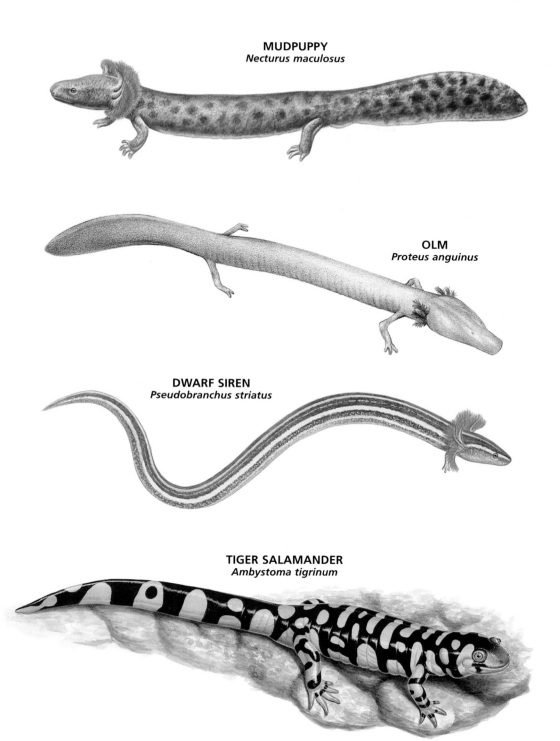

MUDPUPPY
Necturus maculosus

OLM
Proteus anguinus

DWARF SIREN
Pseudobranchus striatus

TIGER SALAMANDER
Ambystoma tigrinum

Salamanders and Newts

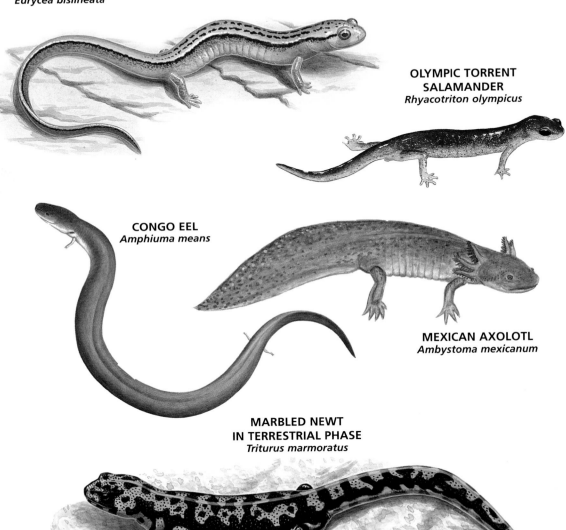

TWO-LINED SALAMANDER
Eurycea bislineata

OLYMPIC TORRENT SALAMANDER
Rhyacotriton olympicus

CONGO EEL
Amphiuma means

MEXICAN AXOLOTL
Ambystoma mexicanum

MARBLED NEWT IN TERRESTRIAL PHASE
Triturus marmoratus

JAPANESE GIANT SALAMANDER
Andrias japonicus

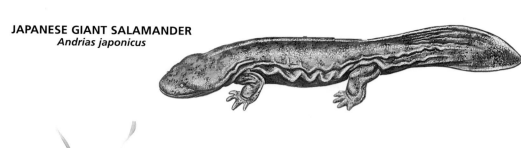

ALPINE NEWT
IN AQUATIC PHASE
Triturus alpestris

RED SALAMANDER
Pseudotriton ruber

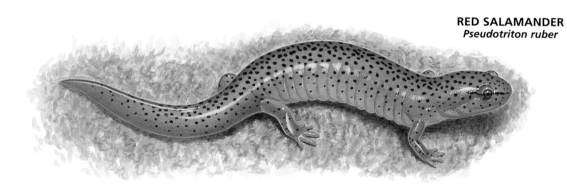

HELLBENDER
Cryptobranchus alleganiensis

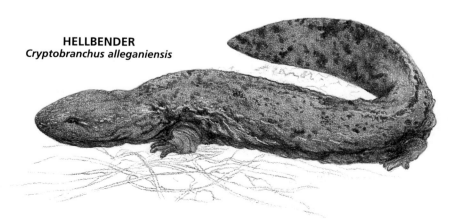

Caecilians

Location

■ Caecilians

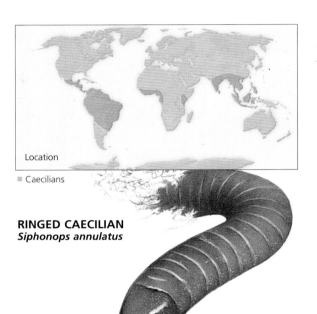

RINGED CAECILIAN
Siphonops annulatus

BANNA CAECILIAN
Ichthyophis bannanicus

Underdeveloped, weak eyes

Caecilians are the only legless amphibians.

Most caecilians live in underground tunnels.

Annuli (rings) partly encircle the body.

ADAPTATIONS FOR BURROWING

Family *Rhinatrematidae*
Epicrionops petersi

Family *Caecilidae*
Microcaecilia rabei

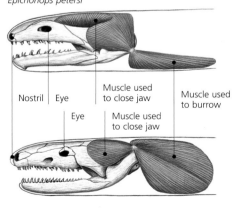

Nostril   Eye

Eye

Muscle used to close jaw

Muscle used to close jaw

Muscle used to burrow

Family *Ichthyophiidae*
Ichthyophis glutinosus

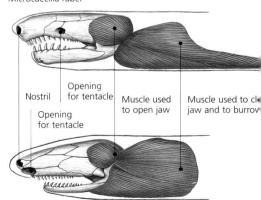

Nostril

Opening for tentacle

Opening for tentacle

Muscle used to open jaw

Muscle used to close jaw and to burrow

Family *Scolecomorphidae*
Crotaphatrema lamottei

AQUATIC CAECILIAN
Typhlonectes natans

SOUTHEAST ASIAN CAECILIAN
Ichthyophis kohtaoensis

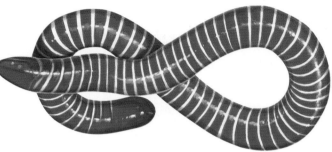

TERRESTRIAL CAECILIAN
Dermophis mexicanus

SAO TOMÉ CAECILIAN
Schistometopum thomense

CAYENNE CAECILIAN
Typhlonectes compressicauda

Frogs and Toads

EVOLUTION OF FROGS AND TOADS

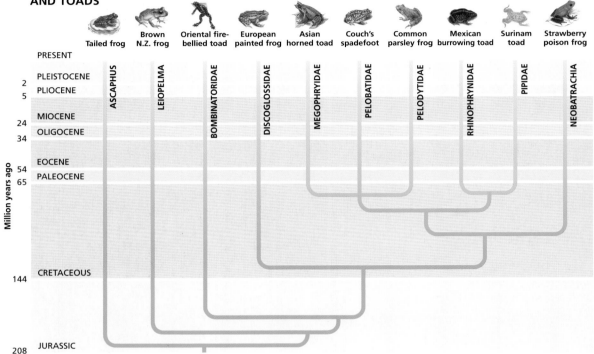

Tailed frog | Brown N.Z. frog | Oriental fire-bellied toad | European painted frog | Asian horned toad | Couch's spadefoot | Common parsley frog | Mexican burrowing toad | Surinam toad | Strawberry poison frog

TREEFROG

TREEFROG SKELETON

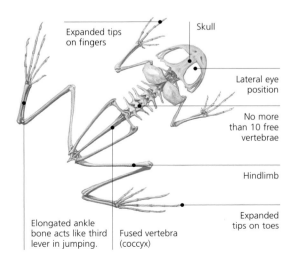

Expanded tips on fingers

Skull

Lateral eye position

No more than 10 free vertebrae

Hindlimb

Expanded tips on toes

Elongated ankle bone acts like third lever in jumping.

Fused vertebra (coccyx)

FROG LIFECYCLE

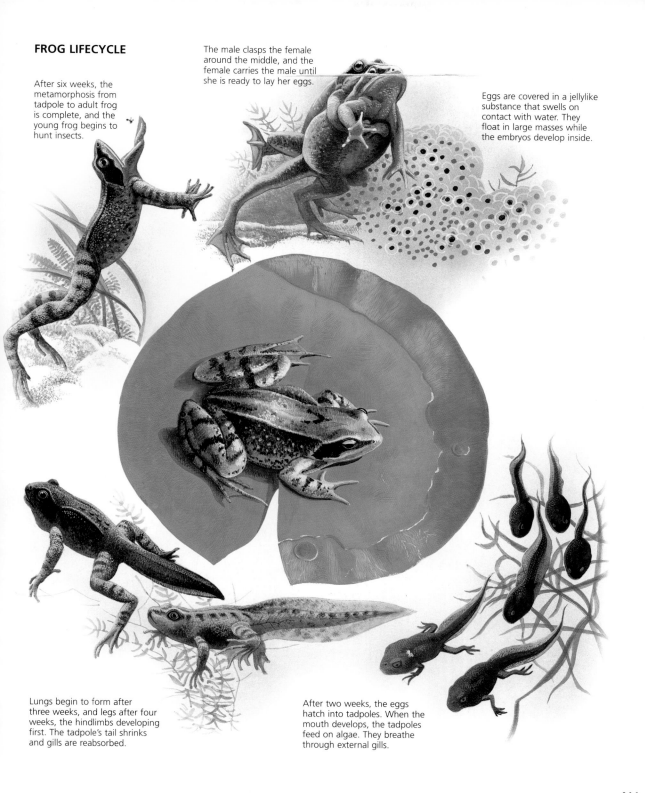

After six weeks, the metamorphosis from tadpole to adult frog is complete, and the young frog begins to hunt insects.

The male clasps the female around the middle, and the female carries the male until she is ready to lay her eggs.

Eggs are covered in a jellylike substance that swells on contact with water. They float in large masses while the embryos develop inside.

Lungs begin to form after three weeks, and legs after four weeks, the hindlimbs developing first. The tadpole's tail shrinks and gills are reabsorbed.

After two weeks, the eggs hatch into tadpoles. When the mouth develops, the tadpoles feed on algae. They breathe through external gills.

Primitive Frogs

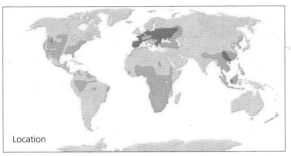

Location

- Primitive frogs
- Pipids and transitional frogs
- Primitive frogs, pipids and transitional frogs

YELLOW-BELLIED TOAD
Bombina variegata

SURINAM TOAD
Pipa pipa

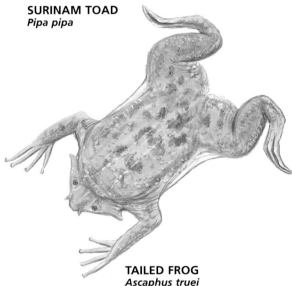

ASIAN HORNED TOAD
Megophrys nasuta

TAILED FROG
Ascaphus truei

MEXICAN BURROWING TOAD
Rhinophrynus dorsalis

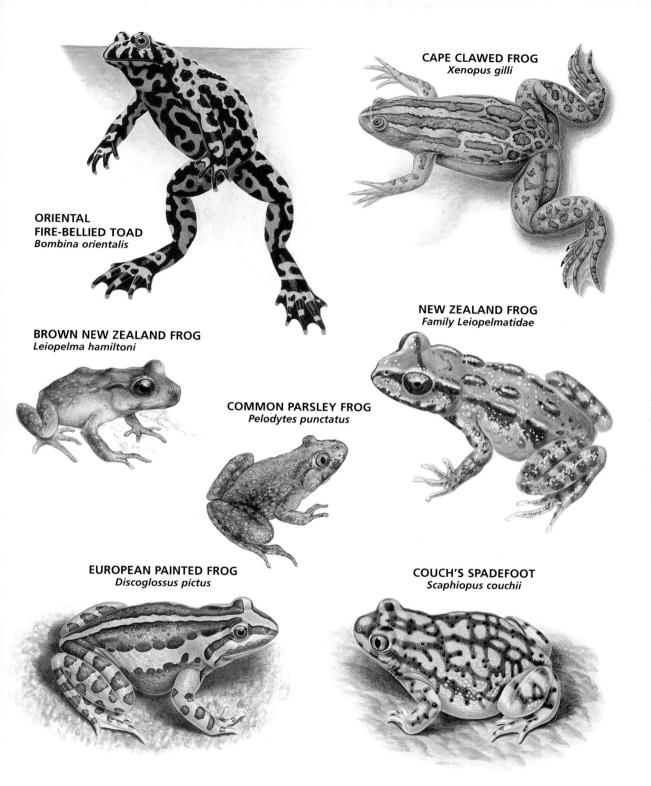

**ORIENTAL
FIRE-BELLIED TOAD**
Bombina orientalis

CAPE CLAWED FROG
Xenopus gilli

NEW ZEALAND FROG
Family Leiopelmatidae

BROWN NEW ZEALAND FROG
Leiopelma hamiltoni

COMMON PARSLEY FROG
Pelodytes punctatus

EUROPEAN PAINTED FROG
Discoglossus pictus

COUCH'S SPADEFOOT
Scaphiopus couchii

413

American and Australasian Frogs

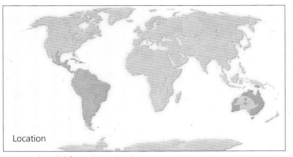

Location

- Leptodactylid frogs (American)
- Myobatrachid frogs (Australasian)

SOUTHERN PLATYPUS FROG
Rheobatrachus silus

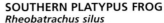

GIANT BULLFROG
Limnodynastes interioris

CRUCIFIX TOAD
Notaden bennettii

WESTERN BARKING FROG
Eleutherodactylus augusti

WESTERN MARSH FROG
Heleioporus barycragus

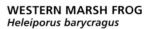

CORROBOREE FROG
Pseudophryne corroboree

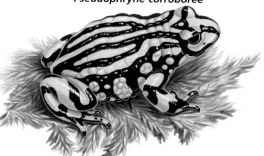

SPOTTED GRASS FROG
Limnodynastes tasmaniensis

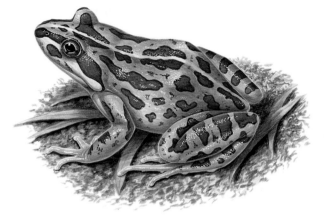

ORNATE HORNED TOAD
Ceratophrys ornata

SOUTH AMERICAN BULLFROG
Leptodactylus pentadactylus

SCHMIDT'S FOREST FROG
Hydrolaetare schmidti

Toads, Treefrogs and Glass Frogs

Location

- Toads and harlequin frogs
- Treefrogs and glass frogs
- Toads, harlequin frogs, treefrogs and glass frogs

**EVERETT'S ASIAN
TREE TOAD**
Pedostibes everetti

VARIABLE HARLEQUIN FROG
Atelopus varius

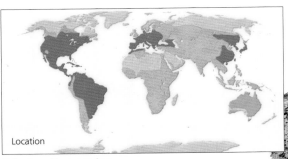

LEOPARD TOAD
Bufo pardalis

**GREEN AND GOLDEN
BELL FROG**
Litoria aurea

ASIATIC CLIMBING TOAD
Pedostibes hosii

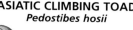

DAINTY GREEN TREEFROG
Litoria gracilenta

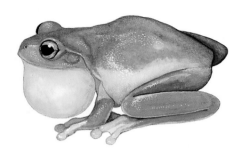

GLASS FROG
Hylinobatrachium fleischmanni

BURROWING TREEFROG
Pternohyla fodiens

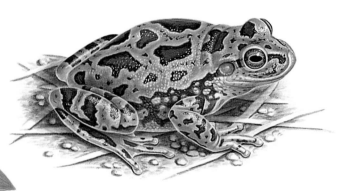

EUROPEAN GREEN TOAD
Bufo viridis

RED-EYED TREEFROG
Agalychnis callidryas

True Frogs and Poison Frogs

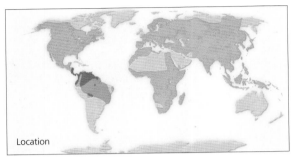

Location

- True frogs
- Poison frogs
- True frogs and poison frogs

ORNATE BURROWING FROG
Hildebrandtia ornata

FUNEREAL POISON FROG
Phyllobates lugubris

SOLOMON ISLANDS TREEFROG
Platymantis guppyi

NORTHERN LEOPARD FROG
Rana sp.

ORANGE AND BLACK POISON FROG
Dendrobates leucomelas

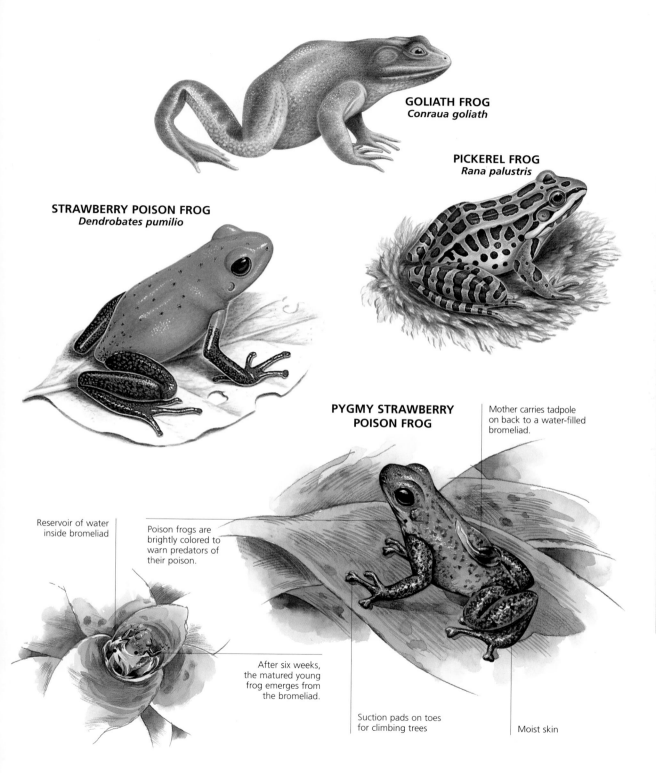

GOLIATH FROG
Conraua goliath

PICKEREL FROG
Rana palustris

STRAWBERRY POISON FROG
Dendrobates pumilio

PYGMY STRAWBERRY POISON FROG

Mother carries tadpole on back to a water-filled bromeliad.

Reservoir of water inside bromeliad

Poison frogs are brightly colored to warn predators of their poison.

After six weeks, the matured young frog emerges from the bromeliad.

Suction pads on toes for climbing trees

Moist skin

419

Other Frogs and Toads

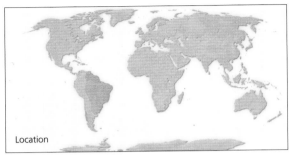

Location

■ Other frogs and toads

GOLD FROG
Brachycephalus ephippium

CAPE GHOST FROG
Heleophryne purcelli

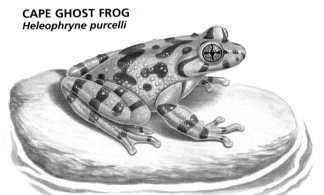

SEYCHELLES FROG
Sooglossus sechellensis

ASIAN PAINTED FROG
Kaloula pulchra

SENEGAL RUNNING FROG
Kassina senegalensis

DARWIN'S FROG
Rhinoderma darwinii

WALLACE'S FLYING FROG
Rhacophorus nigropalmatus

EASTERN
NARROW-MOUTHED TOAD
Gastrophryne carolinensis

PARADOX FROG
Pseudis paradoxa

PAINTED REED FROG
Hyperolius marmoratus

RED-BANDED CREVICE
CREEPER
Phrynomerus bifasciatus

Classifying Fishes

EVOLUTION OF FISHES

European brook lamprey
Leafy seadragon
Atlantic cod
Opah
Variegated lizardfish
Black dragonfish
Northern pike
Atlantic salmon
Striped catfish
South American knifefish

Million years ago		HYPEROARTIA	ACANTHOPTERYGII	PARACANTHOPTERYGII	SCOPELOMORPHA	CYCLOSQUAMATA	STENOPTERYGII	ESOCIFORMES	SALMONIFORMES	SILURIFORMES	GYMNOTIFORMES
2	PRESENT PLEISTOCENE PLIOCENE										
5											
24	MIOCENE										
34	OLIGOCENE										
54	EOCENE										
65	PALEOCENE										
144	CRETACEOUS										
208	JURASSIC										
248	TRIASSIC										
286	PERMIAN										
360	CARBONIFEROUS										
408	DEVONIAN										
435	SILURIAN										
505	ORDOVICIAN										

Barramundi

Flyingfish

Seahorse

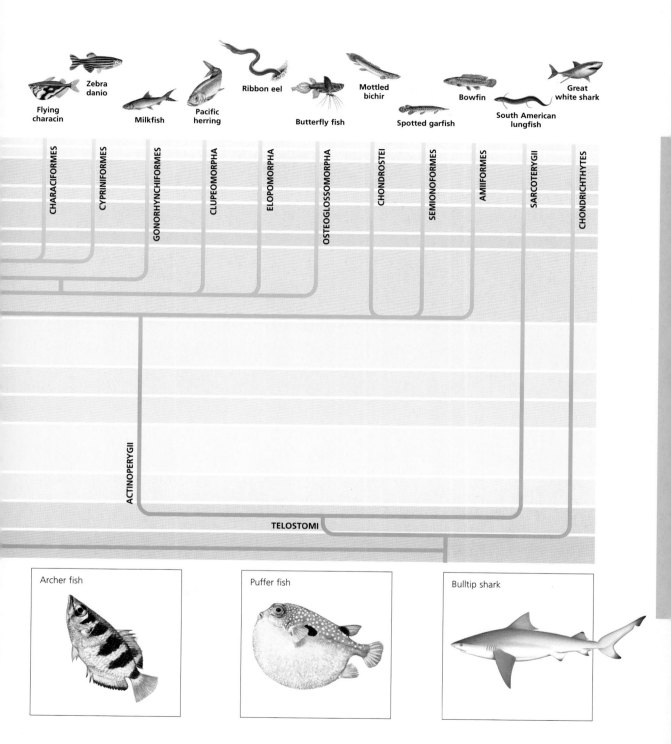

Zebra danio

Flying characin

Milkfish

Pacific herring

Ribbon eel

Butterfly fish

Mottled bichir

Spotted garfish

Bowfin

South American lungfish

Great white shark

CHARACIFORMES

CYPRINIFORMES

GONORHYNCHIFORMES

CLUPEOMORPHA

ELOPOMORPHA

OSTEOGLOSSOMORPHA

CHONDROSTEI

SEMIONOFORMES

AMIIFORMES

SARCOTERYGII

CHONDRICHTHYES

ACTINOPERYGII

TELOSTOMI

Archer fish

Puffer fish

Bulltip shark

Fish Characteristics

BONY FISH

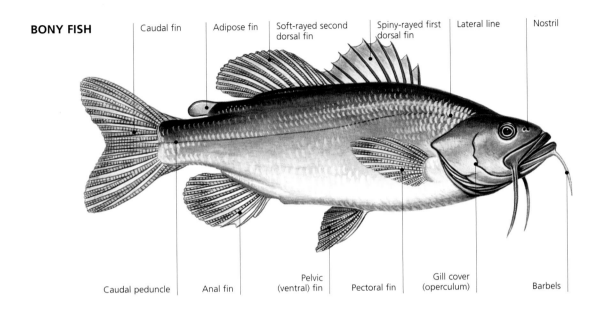

Caudal fin

Adipose fin

Soft-rayed second dorsal fin

Spiny-rayed first dorsal fin

Lateral line

Nostril

Caudal peduncle

Anal fin

Pelvic (ventral) fin

Pectoral fin

Gill cover (operculum)

Barbels

CARTILAGINOUS FISH

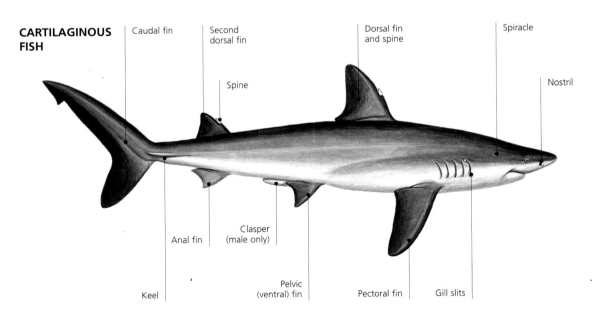

Caudal fin

Second dorsal fin

Dorsal fin and spine

Spiracle

Spine

Nostril

Keel

Anal fin

Clasper (male only)

Pelvic (ventral) fin

Pectoral fin

Gill slits

FISH SCALES

Ctenoid: bony fishes

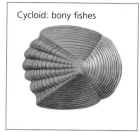
Cycloid: bony fishes

Ganoid: armored fishes

Placoid: sharks

FISH RESPIRATION

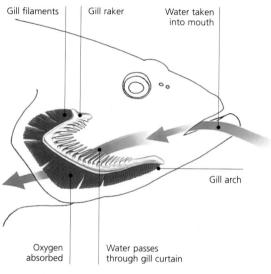

Gill filaments

Gill raker

Water taken into mouth

Gill arch

Oxygen absorbed

Water passes through gill curtain

SHARK RESPIRATION

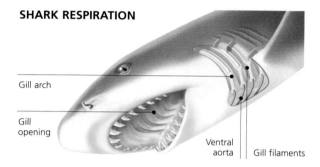

Gill arch

Gill opening

Ventral aorta

Gill filaments

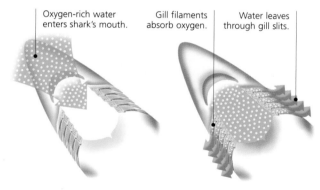

Oxygen-rich water enters shark's mouth.

Gill filaments absorb oxygen.

Water leaves through gill slits.

FISH TAILS

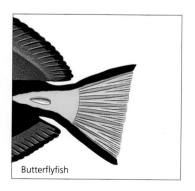
Butterflyfish

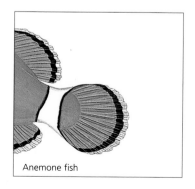
Anemone fish

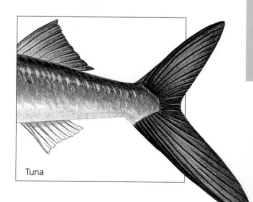
Tuna

Evolution of Fishes

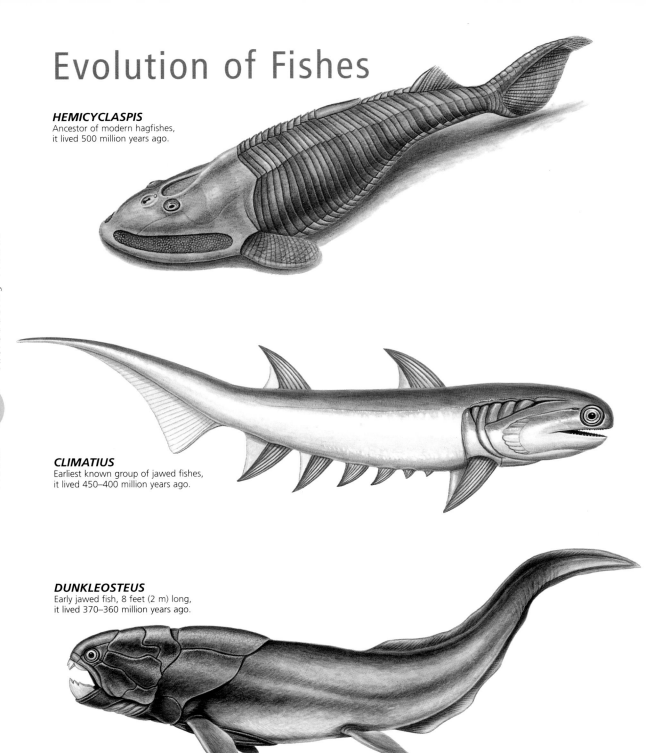

HEMICYCLASPIS
Ancestor of modern hagfishes,
it lived 500 million years ago.

CLIMATIUS
Earliest known group of jawed fishes,
it lived 450–400 million years ago.

DUNKLEOSTEUS
Early jawed fish, 8 feet (2 m) long,
it lived 370–360 million years ago.

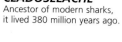

CLADOSELACHE
Ancestor of modern sharks,
it lived 380 million years ago.

CHEIROLEPIS
Possibly an ancestor of sturgeons and
bichirs, it lived 380 million years ago.

EUSTHENOPTERON
Lobe-finned fish with simple lungs,
it lived 360 million years ago.

Jawless Fishes

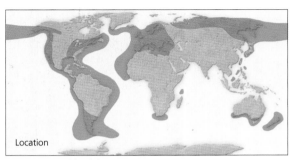

Location

■ Lampreys and hagfishes

SEA LAMPREY
Petromyzon marinus

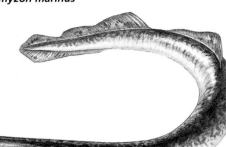

EUROPEAN BROOK LAMPREY
Lampetra planeri

EUROPEAN RIVER LAMPREY
Lampetra fluviatilis

ATLANTIC HAGFISH
Myxine glutinosa

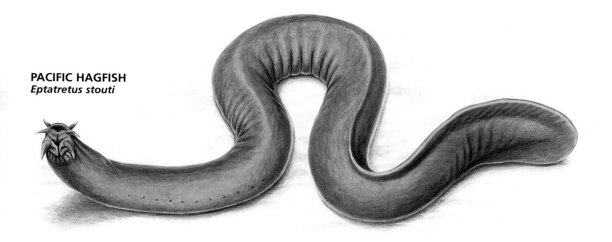

PACIFIC HAGFISH
Eptatretus stouti

HAGFISH KNOTS
By tying themselves in a knot, and bracing themselves against a carcass, limbless and jawless hagfishes gain enough leverage to tear off pieces of flesh.

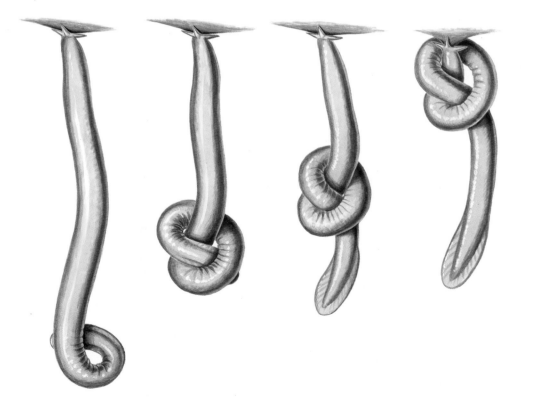

Sharks

EVOLUTION OF SHARKS

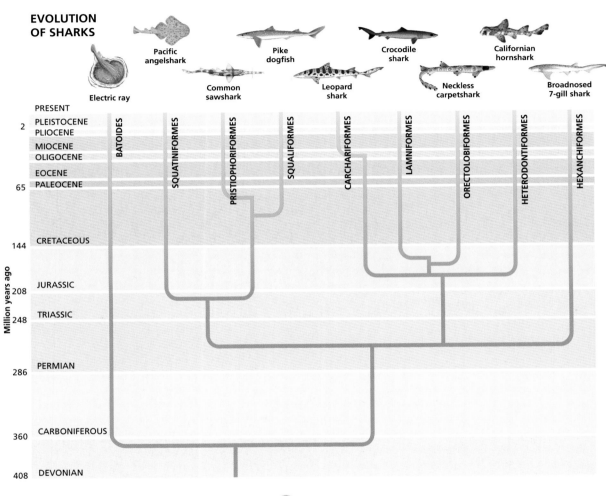

Pacific angelshark

Pike dogfish

Crocodile shark

Californian hornshark

Electric ray

Common sawshark

Leopard shark

Neckless carpetshark

Broadnosed 7-gill shark

Million years ago

PRESENT	
2	PLEISTOCENE
	PLIOCENE
	MIOCENE
	OLIGOCENE
	EOCENE
65	PALEOCENE
144	CRETACEOUS
208	JURASSIC
248	TRIASSIC
286	PERMIAN
360	CARBONIFEROUS
408	DEVONIAN

BATOIDES

SQUATINIFORMES

PRISTIOPHORIFORMES

SQUALIFORMES

CARCHARIFORMES

LAMNIFORMES

ORECTOLOBIFORMES

HETERODONTIFORMES

HEXANCHIFORMES

ELECTROSENSE

Sharks are able to detect electricity in seawater. They use this unique sense to locate prey. Electrosensory organs, known as ampullae of Lorenzini, are spread across a shark's snout.

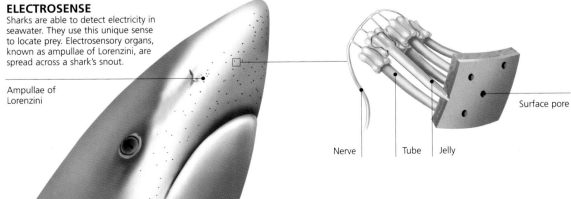

Ampullae of Lorenzini

Nerve   Tube   Jelly

Surface pore

SHARK REPRODUCTION

Hornsharks
Hornsharks lay oval egg cases ringed with screwlike ridges. The mother wedges each case firmly into a crevice.

Swellsharks
Swellsharks lay flat, rectangular egg cases.

A growing shark and the yolk that feeds it.

The newborn shark begins to emerge.

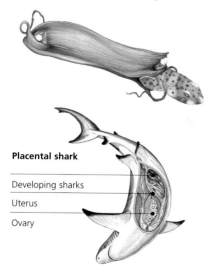

Placental shark

Developing sharks

Uterus

Ovary

TYPES OF TEETH

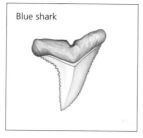

Blue shark

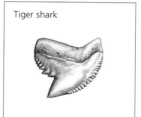

Tiger shark

Hornshark

Great white shark

Shortfin mako

ANATOMY OF FEEDING

In its resting position, the shark's jaw lies just under its brain case.

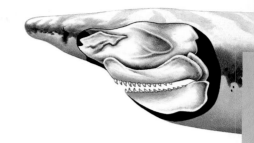

As the shark goes to grab its prey, the upper jaw slides forward. The lower jaw drops.

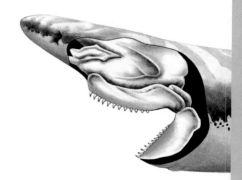

Muscle contractions force the jaw out of the mouth to give the shark a good grip.

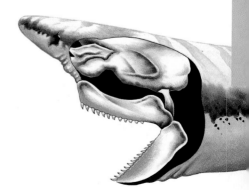

Angel Sharks, Sawfishes and Dogfish Sharks

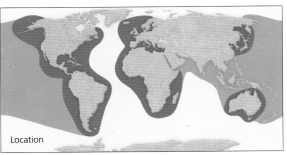

Location

- Sawfishes and dogfish sharks
- Sawfishes, dogfish sharks and angel sharks

SPINED PYGMY SHARK
Squaliolus laticaudus

COOKIE-CUTTER SHARK
Isistius brasiliensis

SMALLFIN GULPER SHARK
Centrophorus moluccensis

SPINY DOGFISH
Squalus acanthias

Sharks

FISHES

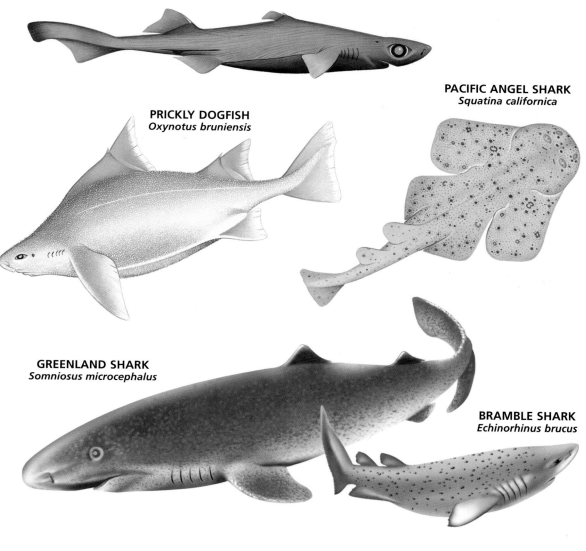

BLACKBELLY LANTERNSHARK
Etmopterus lucifer

PACIFIC ANGEL SHARK
Squatina californica

PRICKLY DOGFISH
Oxynotus bruniensis

GREENLAND SHARK
Somniosus microcephalus

BRAMBLE SHARK
Echinorhinus brucus

LONGNOSE SAWFISH
Pristiophorus cirratus

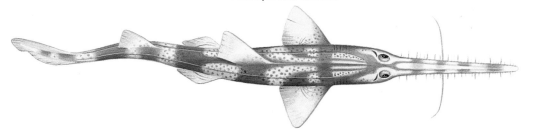

Ground Sharks

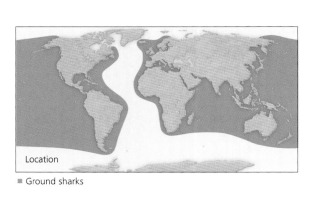

Location

■ Ground sharks

TYPES OF TAILS

Gray reef shark

LEOPARD SHARK
Triakis semifasciata

Oceanic whitetip shark

BLACKTIP REEF SHARK
Carcharhinus melanopterus

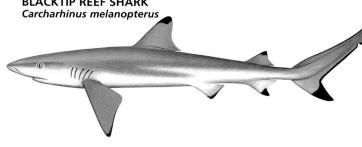

Tiger shark

Atlantic weasel shark

GRACEFUL CATSHARK
Proscyllium habereri

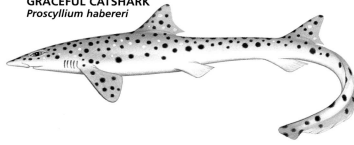

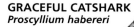

Swell shark

SCALLOPED HAMMERHEAD
Sphyrna lewini

HEAD SHAPES

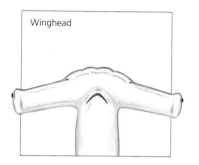

Winghead

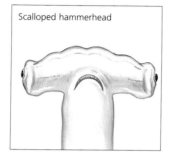

Scalloped hammerhead

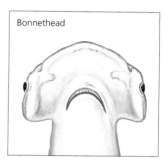

Bonnethead

HEAD HUNTING

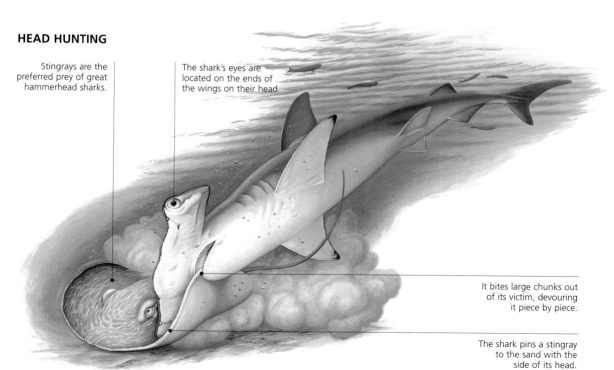

Stingrays are the preferred prey of great hammerhead sharks.

The shark's eyes are located on the ends of the wings on their head.

It bites large chunks out of its victim, devouring it piece by piece.

The shark pins a stingray to the sand with the side of its head.

Mackerel Sharks

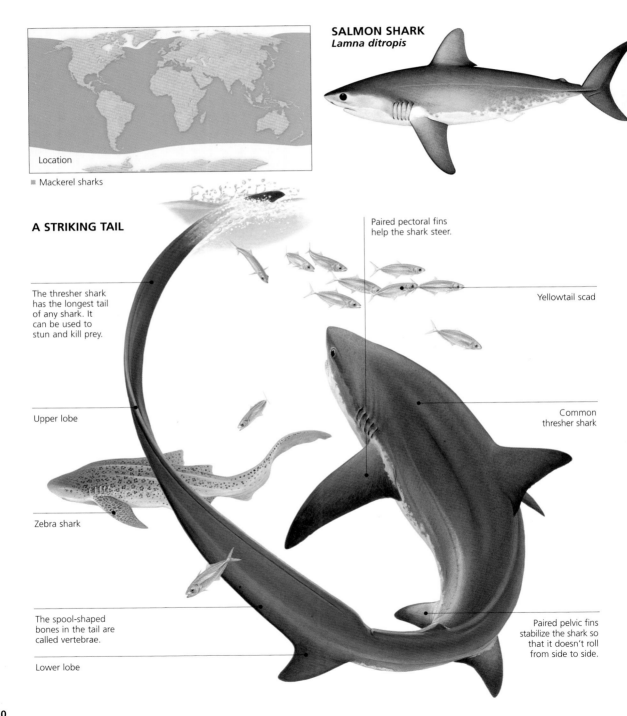

Location

■ Mackerel sharks

SALMON SHARK
Lamna ditropis

A STRIKING TAIL

The thresher shark has the longest tail of any shark. It can be used to stun and kill prey.

Upper lobe

Zebra shark

The spool-shaped bones in the tail are called vertebrae.

Lower lobe

Paired pectoral fins help the shark steer.

Yellowtail scad

Common thresher shark

Paired pelvic fins stabilize the shark so that it doesn't roll from side to side.

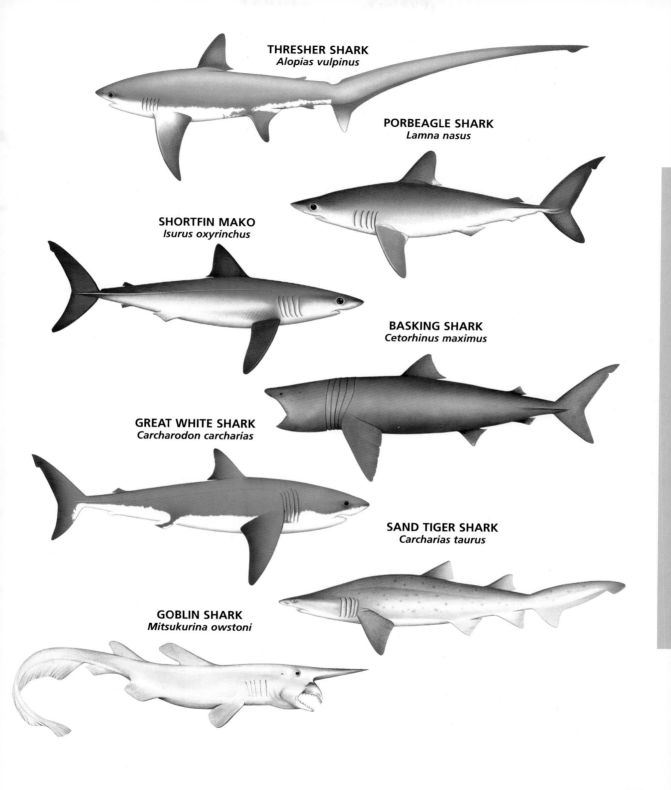

THRESHER SHARK
Alopias vulpinus

PORBEAGLE SHARK
Lamna nasus

SHORTFIN MAKO
Isurus oxyrinchus

BASKING SHARK
Cetorhinus maximus

GREAT WHITE SHARK
Carcharodon carcharias

SAND TIGER SHARK
Carcharias taurus

GOBLIN SHARK
Mitsukurina owstoni

Carpetsharks, Bullhead Sharks, Frilled Sharks and Cow Sharks

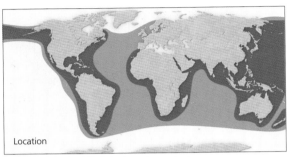

Location

■ Bullhead sharks, frilled sharks and cow sharks
■ Bullhead sharks, frilled sharks, cow sharks and carpetsharks

WHALE SHARK FILTER FEEDING

Gill slits filter food from water.

Shark swallows filtered food.

Plankton, small fishes and water

CAMOUFLAGE

Some sharks have patterns on their skin that allow them to blend in with their environments. This helps conceal them from prey and predators.

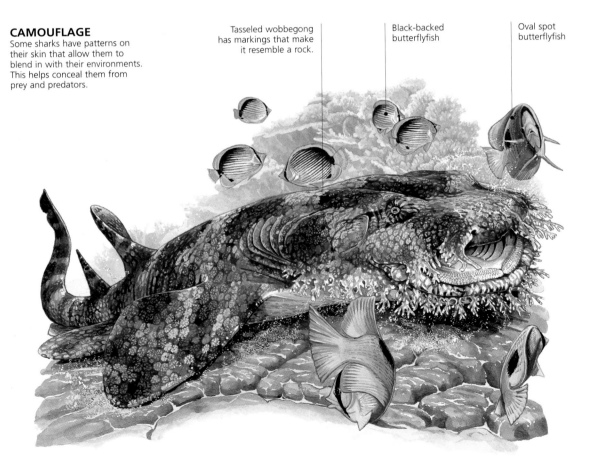

Tasseled wobbegong has markings that make it resemble a rock.

Black-backed butterflyfish

Oval spot butterflyfish

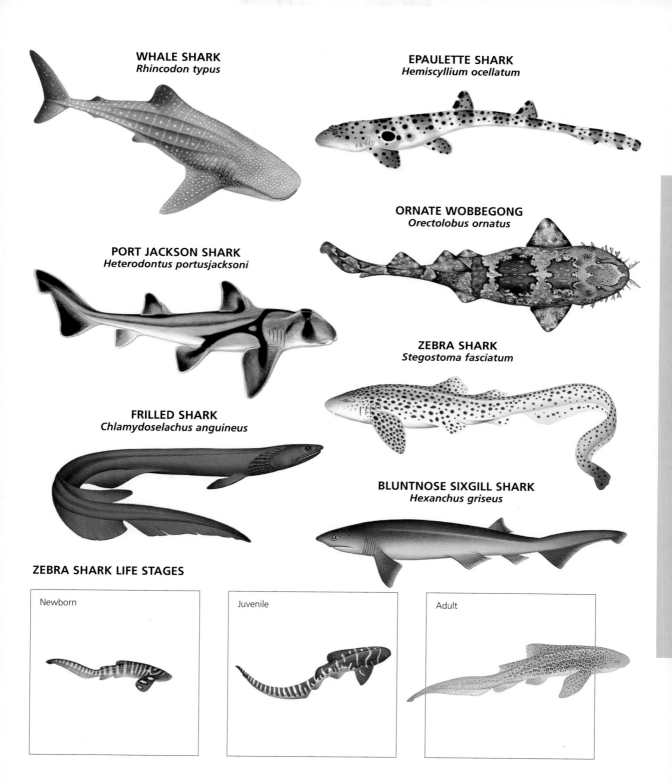

WHALE SHARK
Rhincodon typus

EPAULETTE SHARK
Hemiscyllium ocellatum

ORNATE WOBBEGONG
Orectolobus ornatus

PORT JACKSON SHARK
Heterodontus portusjacksoni

ZEBRA SHARK
Stegostoma fasciatum

FRILLED SHARK
Chlamydoselachus anguineus

BLUNTNOSE SIXGILL SHARK
Hexanchus griseus

ZEBRA SHARK LIFE STAGES

Newborn

Juvenile

Adult

Rays and Chimaeras

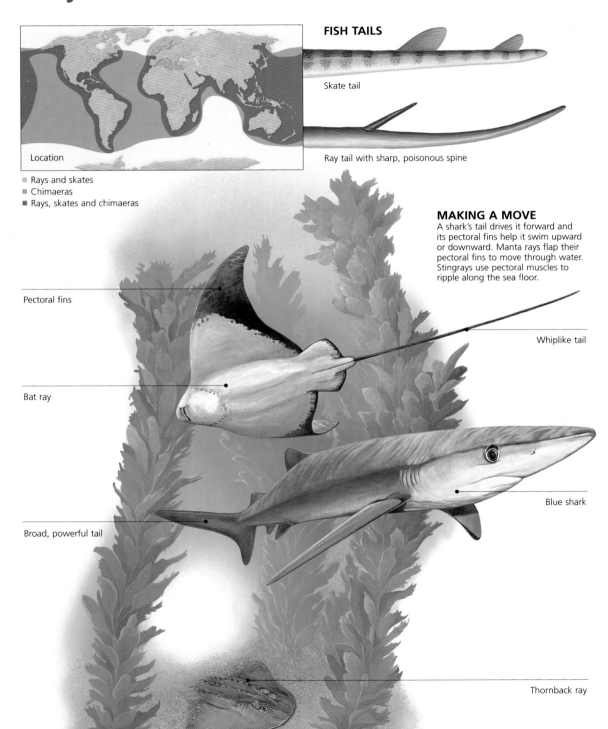

FISH TAILS

Skate tail

Ray tail with sharp, poisonous spine

Location

- Rays and skates
- Chimaeras
- Rays, skates and chimaeras

MAKING A MOVE
A shark's tail drives it forward and its pectoral fins help it swim upward or downward. Manta rays flap their pectoral fins to move through water. Stingrays use pectoral muscles to ripple along the sea floor.

Pectoral fins

Whiplike tail

Bat ray

Blue shark

Broad, powerful tail

Thornback ray

Spookfish

Atlantic devil ray

Freshwater sawfish snout

Shortnose chimaera

Butterfly ray tooth
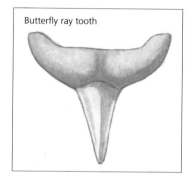

Skate

Elephantfish

Stingray

Guitarfish

Electric ray
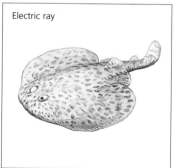

Electric torpedo ray
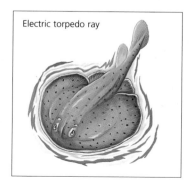

Mouth and nostrils of Atlantic torpedo ray
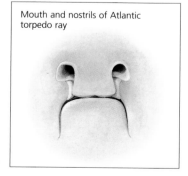

Rays and Chimaeras

BLUNT-NOSED CHIMAERA
Hydrolagus colliei

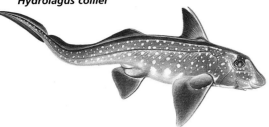

OCELLATED FRESHWATER STINGRAY
Potamotrygon motoro

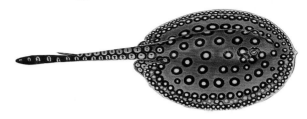

SIXGILL STINGRAY
Hexatrygon bickelli

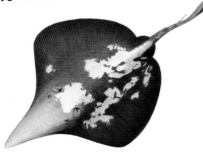

BLIND ELECTRIC RAY
Typhlonarke aysoni

BLUE SKATE
Notoraja sp.

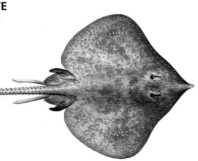

PORT DAVEY SKATE
Raja sp.

HOW THEY SWIM

Birds can overcome gravity because they don't weigh much for their size. The same is true of cartilaginous fishes. Cartilage is lighter than bone, and this allows sharks and rays to slice through the water with minimum effort.

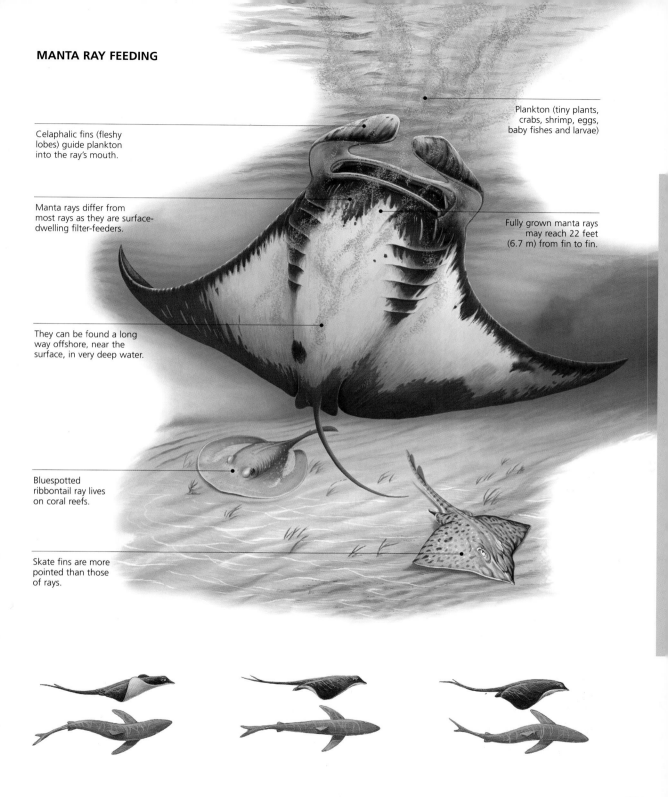

MANTA RAY FEEDING

Plankton (tiny plants, crabs, shrimp, eggs, baby fishes and larvae)

Celaphalic fins (fleshy lobes) guide plankton into the ray's mouth.

Manta rays differ from most rays as they are surface-dwelling filter-feeders.

Fully grown manta rays may reach 22 feet (6.7 m) from fin to fin.

They can be found a long way offshore, near the surface, in very deep water.

Bluespotted ribbontail ray lives on coral reefs.

Skate fins are more pointed than those of rays.

Bony Fishes

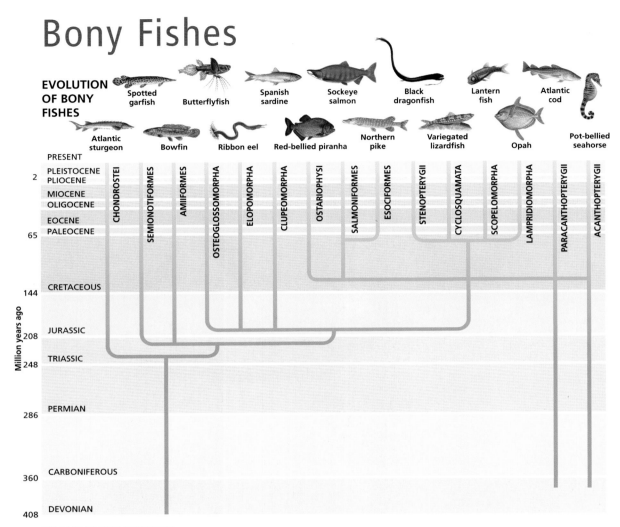

EVOLUTION OF BONY FISHES

Spotted garfish

Butterflyfish

Spanish sardine

Sockeye salmon

Black dragonfish

Lantern fish

Atlantic cod

Atlantic sturgeon

Bowfin

Ribbon eel

Red-bellied piranha

Northern pike

Variegated lizardfish

Opah

Pot-bellied seahorse

PRESENT

2 PLEISTOCENE PLIOCENE

MIOCENE

OLIGOCENE

EOCENE

65 PALEOCENE

144 CRETACEOUS

208 JURASSIC

248 TRIASSIC

286 PERMIAN

360 CARBONIFEROUS

408 DEVONIAN

Million years ago

CHONDROSTEI

SEMIONOTIFORMES

AMIIFORMES

OSTEOGLOSSOMORPHA

ELOPOMORPHA

CLUPEOMORPHA

OSTARIOPHYSI

SALMONIFORMES

ESOCIFORMES

STENOPTERYGII

CYCLOSQUAMATA

SCOPELOMORPHA

LAMPRIDIOMORPHA

PARACANTHOPTERYGII

ACANTHOPTERYGII

Bony Fishes

FISHES

TROUT REPRODUCTION

Female brown trout lays eggs.

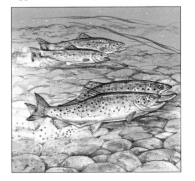

Eggs are fertilized by the male.

Hatchlings feed on orange yolk sac.

448

BLUE-FIN TUNA

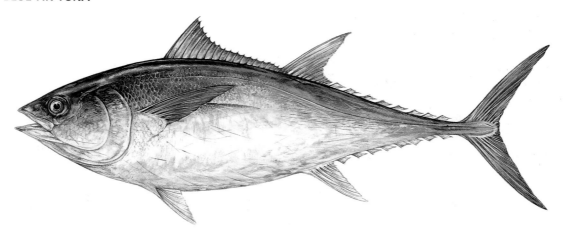

TUNA SKELETON

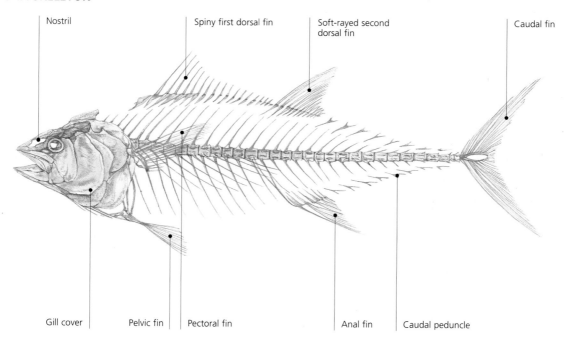

Nostril

Spiny first dorsal fin

Soft-rayed second
dorsal fin

Caudal fin

Gill cover

Pelvic fin

Pectoral fin

Anal fin

Caudal peduncle

Lungfishes and Coelacanth

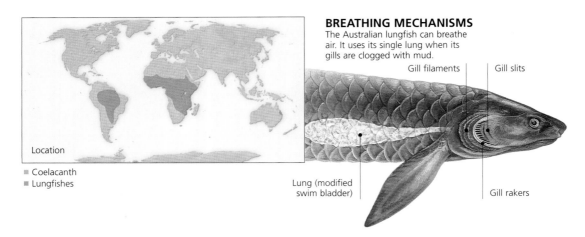

Location

- Coelacanth
- Lungfishes

BREATHING MECHANISMS

The Australian lungfish can breathe air. It uses its single lung when its gills are clogged with mud.

Gill filaments

Gill slits

Lung (modified swim bladder)

Gill rakers

SOUTH AMERICAN LUNGFISH
Lepidosiren paradoxa

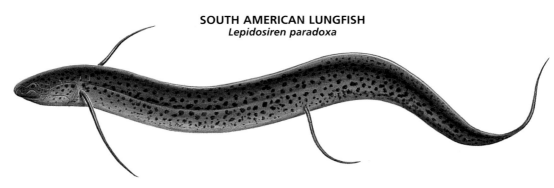

AUSTRALIAN LUNGFISH
Neoceratodus forsteri

AFRICAN LUNGFISH
Protopterus dolloi

COELACANTH
Latimeria chalumnae

CRAWLING IN WATER

The African lungfish can swim and crawl. When crawling, it raises its body and propels itself forward with its thin, paired fins.

Bichirs

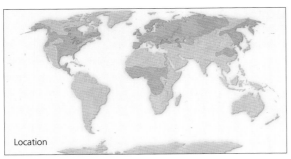

Location

■ Bichirs and relatives

MOTTLED BICHIR
Polypterus weeksi

ATLANTIC STURGEON
Acipenser oxyrinchus

BOWFIN
Amia calva

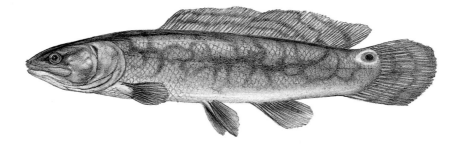

SPOTTED GAR
Lepisosteus oculatus

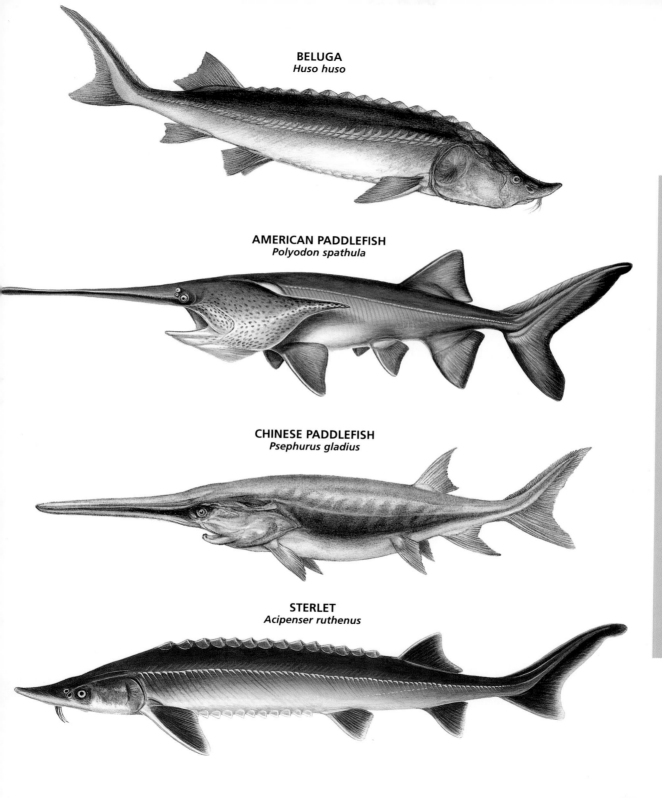

BELUGA
Huso huso

AMERICAN PADDLEFISH
Polyodon spathula

CHINESE PADDLEFISH
Psephurus gladius

STERLET
Acipenser ruthenus

453

Bonytongues

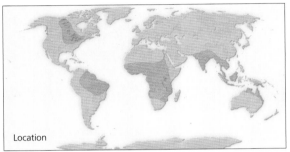

Location

■ Bonytongues and relatives

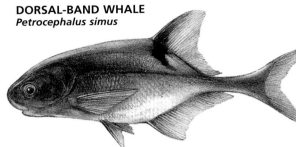

DORSAL-BAND WHALE
Petrocephalus simus

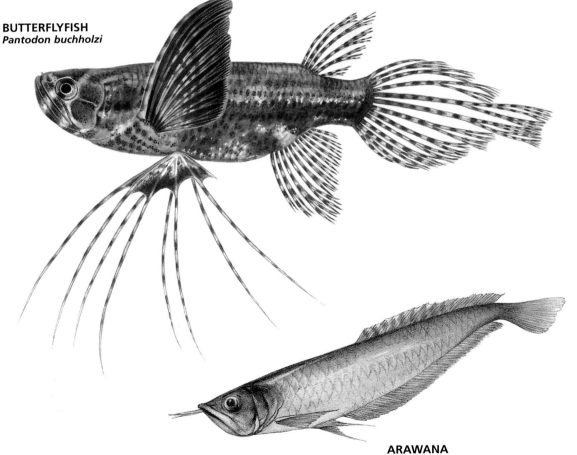

BUTTERFLYFISH
Pantodon buchholzi

ARAWANA
Osteoglossum bicirrhosum

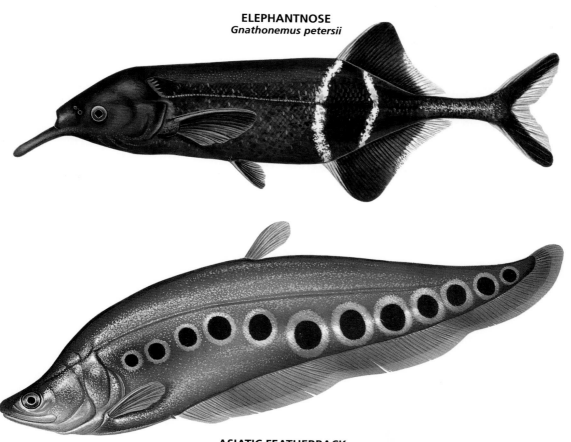

ELEPHANTNOSE
Gnathonemus petersii

ASIATIC FEATHERBACK
Chitala chitala

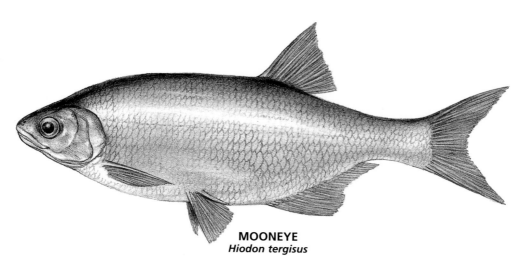

MOONEYE
Hiodon tergisus

Eels

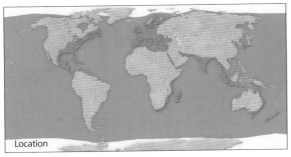

Location

■ Eels and relatives

Silver or mature

AMERICAN EEL
Anguilla rostrata

JUVENILE GARDEN EELS
Being able to burrow tail-first into the shell bottom of tropical seas, garden eels are adapted to life firmly anchored into the sea floor. They emerge from their burrows to feed on plankton.

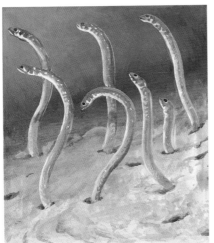

Yellow or juvenile

TESSELLATED MORAY EEL
Gymnothorax favagineus

RIBBON EEL
Rhinomuraena quaesita

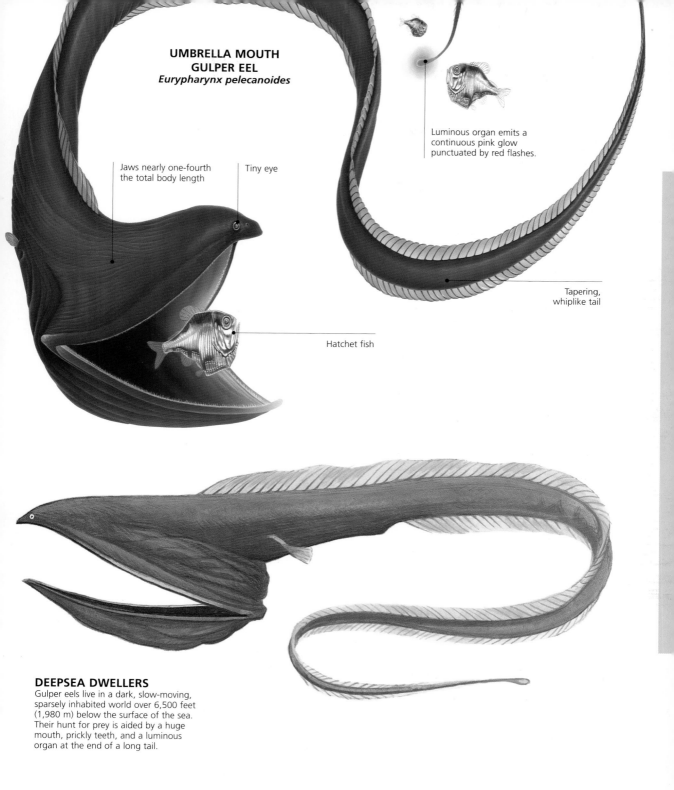

UMBRELLA MOUTH GULPER EEL
Eurypharynx pelecanoides

Jaws nearly one-fourth the total body length

Tiny eye

Luminous organ emits a continuous pink glow punctuated by red flashes.

Hatchet fish

Tapering, whiplike tail

DEEPSEA DWELLERS
Gulper eels live in a dark, slow-moving, sparsely inhabited world over 6,500 feet (1,980 m) below the surface of the sea. Their hunt for prey is aided by a huge mouth, prickly teeth, and a luminous organ at the end of a long tail.

Sardines

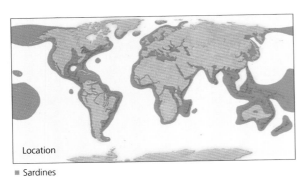

Location

■ Sardines

ATLANTIC THREAD HERRING
Opisthonema oglinum

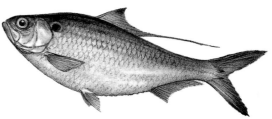

PERUVIAN ANCHOVETA
Engraulis ringens

AMERICAN GIZZARD SHAD
Dorosoma cepedianum

CASPIAN SPRAT
Clupeonella cultriventris

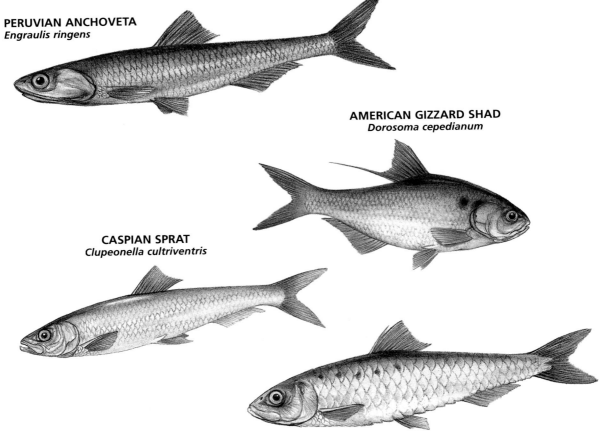

EUROPEAN PILCHARD
Sardina pilchardus

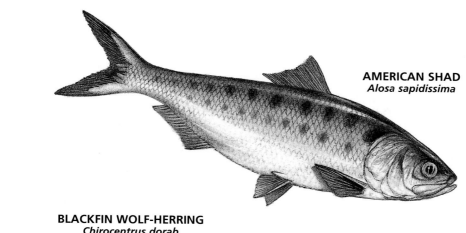

AMERICAN SHAD
Alosa sapidissima

BLACKFIN WOLF-HERRING
Chirocentrus dorab

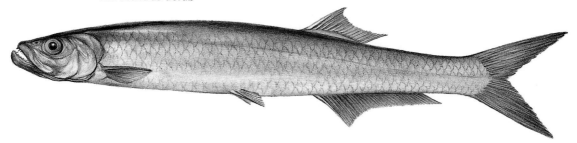

SPANISH SARDINE
Sardinella aurita

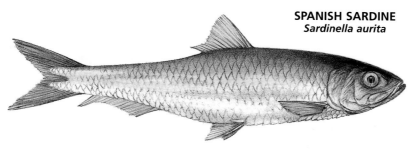

TYPES OF SARDINES

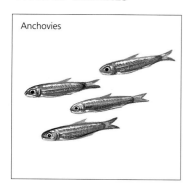

Anchovies

Pacific herring

Sardines

Catfishes, Carps and Characins

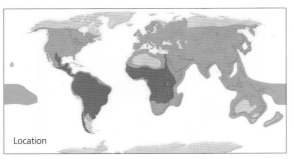

Location

- Charachins
- Catfishes and carps
- Charachins, catfishes and carps

BITTERLING
Rhodeus sericeus

Breeding habits
Female bitterling lays eggs inside a mussel. The male sheds sperm over the mussel's syphon. As the mussel breathes, the sperm is drawn into the cavity to fertilize the eggs.

MILKFISH
Chanos chanos

SLIMY LOACH
Acantophthalmus myersi

EGG INCUBATION
The spotted African squeaker begins life in the mouth of another species of fish. A female cichlid picks up the eggs and broods them in her mouth.

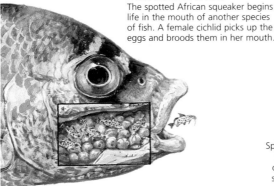

A LEAP FOR LIFE
Splash tetra leap out of the water and spawn on the underside of objects such as leaves. The male splashes water on the eggs with his tail until they hatch.

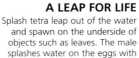

ELECTRIC CATFISH
Malapterurus electricus

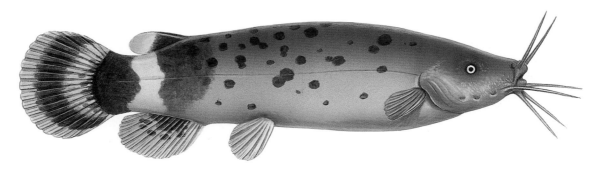

SADDLED HILLSTREAM LOACH
Homaloptera orthogoniata

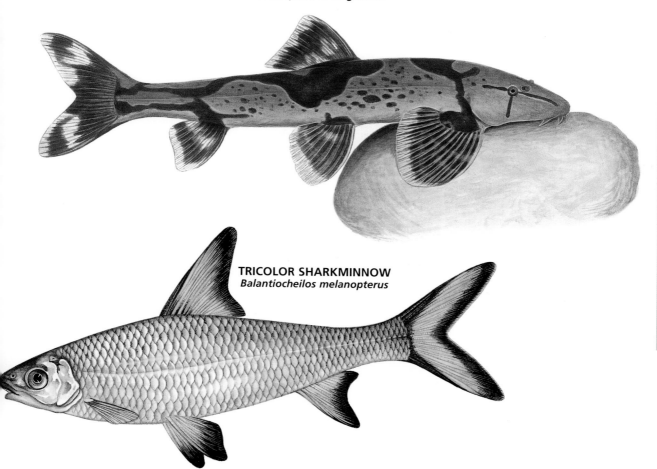

TRICOLOR SHARKMINNOW
Balantiocheilos melanopterus

Catfishes, Carps and Characins

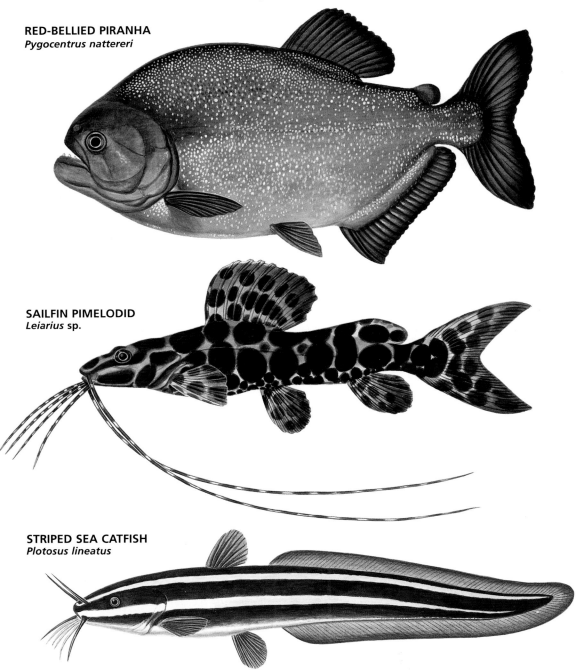

RED-BELLIED PIRANHA
Pygocentrus nattereri

SAILFIN PIMELODID
Leiarius sp.

STRIPED SEA CATFISH
Plotosus lineatus

MARBLED HATCHETFISH
Carnegiella strigata

STRIPED HEADSTANDER
Anostomus anostomus

ZEBRA DANIO
Danio rerio

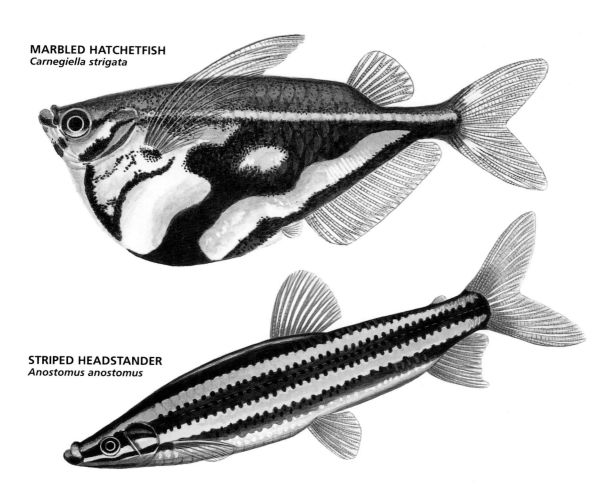

Salmons

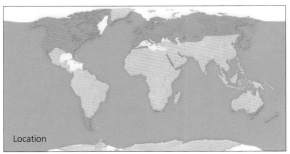

Location

■ Salmons

PIKE HEAD SKELETON

Backward-slanting upper teeth trap prey in pike's mouth.

Large head enables pike to swallow prey almost half its size.

SOCKEYE SALMON LIFE STAGES

Parr lives in rivers and small lakes.

It enters the sea as a smolt.

Smolt matures into a sea-going adult.

ARTHUR'S PARAGALAXIAS
Paragalaxias mesotes

CISCO
Coregonus artedi

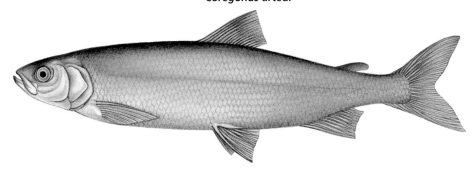

SOCKEYE SALMON LIFECYCLE

Summer–Autumn
Female buries several thousand fertilized eggs in a shallow nest dug in the gravel of a river bed.

Summer–Autumn
During spawning the male turns a brilliant red and develops a long, hooked jaw. After spawning, the salmon soon die.

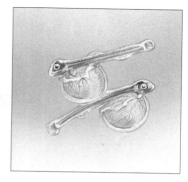

Winter–Spring
Eggs take several months to hatch. Hatchlings remain in the gravel until the attached yolk has been consumed.

Spring
Salmon fry wriggle out into the stream, inhabiting a lake for a year before beginning the journey out to sea.

■ Location

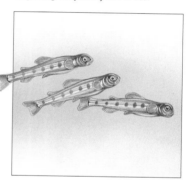

■ Migratory path

Spring
The sea-bound salmon, known as smolts, face many predators such as other fishes, birds and mammals.

Spring–Summer
Returning to their birth river, some are eaten by brown bears. The survivors begin to show dramatic spawning colors.

Spring–Summer
After years at sea, salmon approach sexual maturity and are ready to face the rigors of the spawning migration.

Salmons

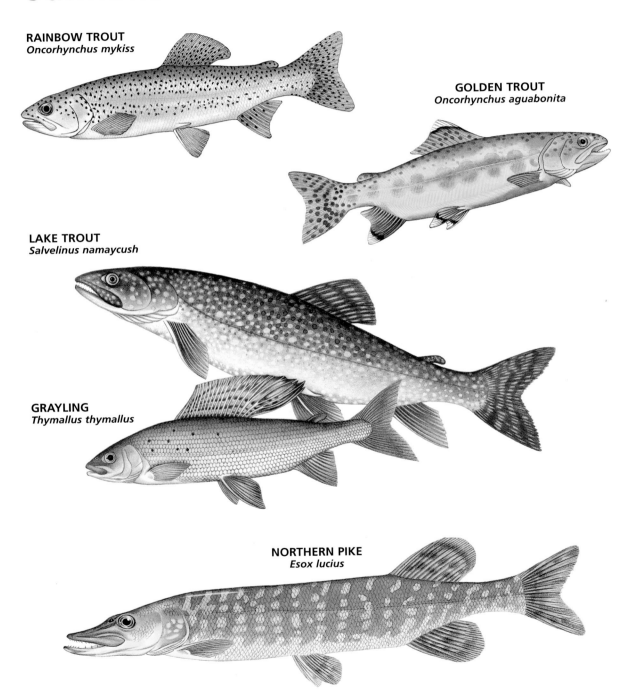

RAINBOW TROUT
Oncorhynchus mykiss

GOLDEN TROUT
Oncorhynchus aguabonita

LAKE TROUT
Salvelinus namaycush

GRAYLING
Thymallus thymallus

NORTHERN PIKE
Esox lucius

ALASKA BLACKFISH
Dallia pectoralis

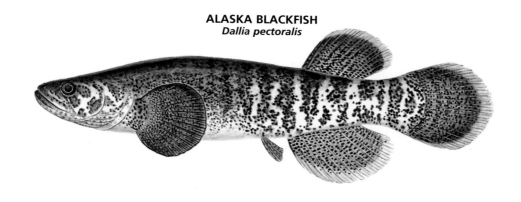

SOCKEYE SALMON
Oncorhynchus nerka

ATLANTIC SALMON
Salmo salar

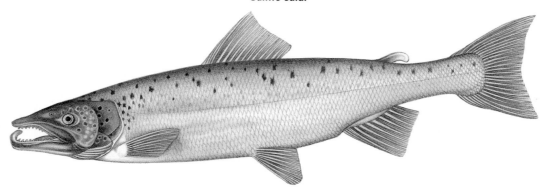

Dragonfishes and Lanternfishes

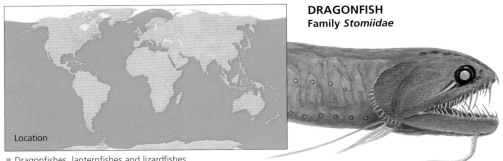

Location

- Dragonfishes, lanternfishes and lizardfishes

DRAGONFISH
Family *Stomiidae*

BRISTLEMOUTH
Family *Gonostomatidae*

CROSS-TOOTHED PERCH
Order *Stomiiformes*

LANTERNFISH
Family *Myctophidae*

VARIEGATED LIZARDFISH
Synodus variegatus

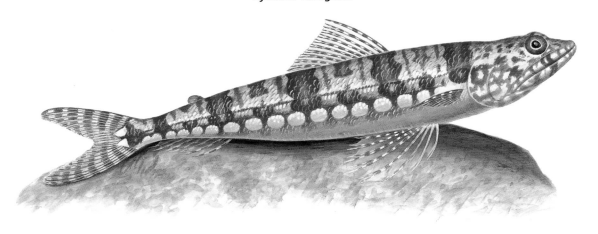

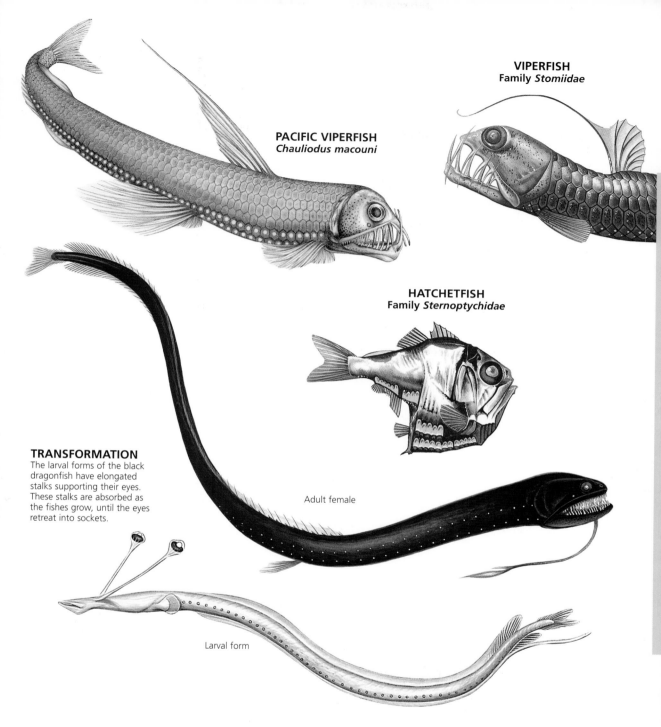

VIPERFISH
Family *Stomiidae*

PACIFIC VIPERFISH
Chauliodus macouni

HATCHETFISH
Family *Sternoptychidae*

TRANSFORMATION
The larval forms of the black dragonfish have elongated stalks supporting their eyes. These stalks are absorbed as the fishes grow, until the eyes retreat into sockets.

Adult female

Larval form

BLACK DRAGONFISH
Idiacanthus fasciola

Cods and Anglerfishes

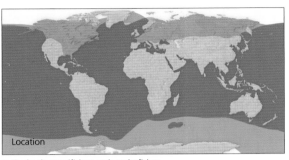

Location

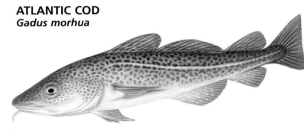

ATLANTIC COD
Gadus morhua

- Cuskeels, toadfishes and anglerfishes
- Cods and troutperches
- Cuskeels, toadfishes, anglerfishes, cods and troutperches

PIRATE PERCH
Aphredoderus sayanus

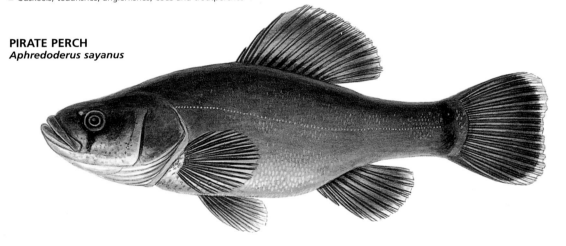

SPLENDID TOADFISH
Sanopus splendidus

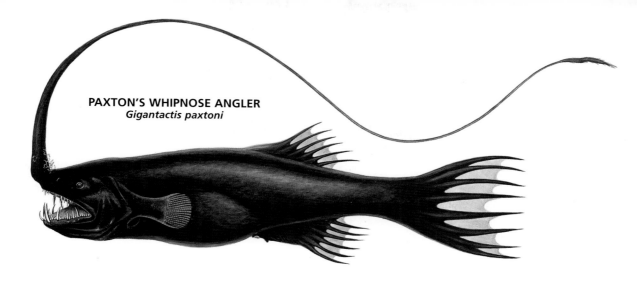

PAXTON'S WHIPNOSE ANGLER
Gigantactis paxtoni

CIRCUMPOLAR BURBOT
Lota lota

PARASITIC LIFESTYLE

Male anglerfishes are tiny compared to females, with toothless jaws and no baits to lure prey. Males sometimes attach themselves to females and live as parasites.

CAMOUFLAGE TECHNIQUES

Bottom-dwelling anglerfishes blend in with their environment to snare prey. Frogfishes can change color and hide among coral.

Roughjaw frogfish

Female

Bullbous bait

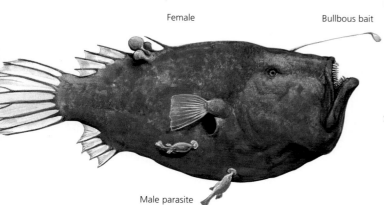

Sargassum fish

Male parasite

Spiny-rayed Fishes

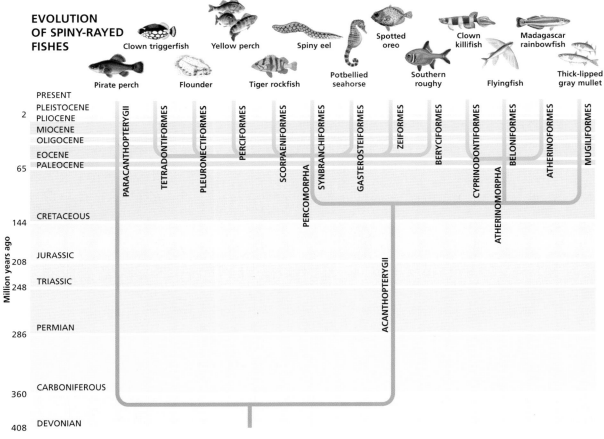

EVOLUTION OF SPINY-RAYED FISHES

Clown triggerfish
Yellow perch
Spiny eel
Spotted oreo
Clown killifish
Madagascar rainbowfish
Pirate perch
Flounder
Tiger rockfish
Potbellied seahorse
Southern roughy
Flyingfish
Thick-lipped gray mullet

PRESENT
PLEISTOCENE
2
PLIOCENE
MIOCENE
OLIGOCENE
EOCENE
PALEOCENE
65
CRETACEOUS
144
JURASSIC
208
TRIASSIC
248
PERMIAN
286
CARBONIFEROUS
360
DEVONIAN
408

Million years ago

PARACANTHOPTERYGII
TETRADONTIFORMES
PLEURONECTIFORMES
PERCIFORMES
SCORPAENIFORMES
PERCOMORPHA
SYNBRANCHIFORMES
GASTEROSTEIFORMES
ZEIFORMES
BERYCIFORMES
ACANTHOPTERYGII
CYPRINODONTIFORMES
ATHERINOMORPHA
BELONIFORMES
ATHERINOFORMES
MUGILIFORMES

STICKLEBACK COURTSHIP

Male uses his bright colors to attract a passing female.

The pair begins a courtship dance.

Male entices female to a nest he has built.

AUSTRALIAN RAINBOWFISH

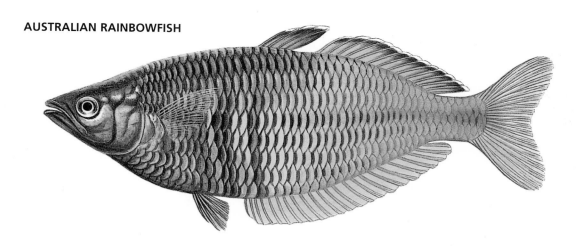

RAINBOWFISH SKELETON

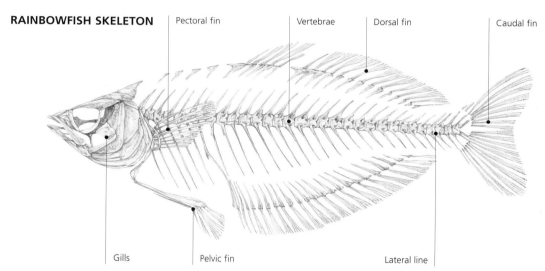

Pectoral fin

Vertebrae

Dorsal fin

Caudal fin

Gills

Pelvic fin

Lateral line

He guides and coaxes her inside.

Once inside, the female lays her eggs, then leaves.

Male fertilizes eggs and guards them until they hatch.

Clingfishes, Flyingfishes, Killifishes, Ricefishes and Silversides

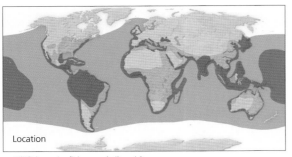

Location

- Killifishes, ricefishes and silversides
- Clingfishes and flyingfishes
- Killifishes, ricefishes, silversides, clingfishes and flyingfishes

MANDARIN FISH
Synchiropus splendidus

SWORDTAIL
Xiphophorus helleri

Changing sexes
The male swordtail is smaller than the female, and has a swordlike extension to its tail. Females sometimes turn into males, even after giving birth.

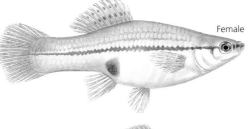

Female

Male

URCHIN CLINGFISH
Diademichthys lineatus

FLYINGFISH
Family *Exocoetidae*

MALAYAN HALFBEAK
Dermogenys pusilla

Bony Fishes

FISHES

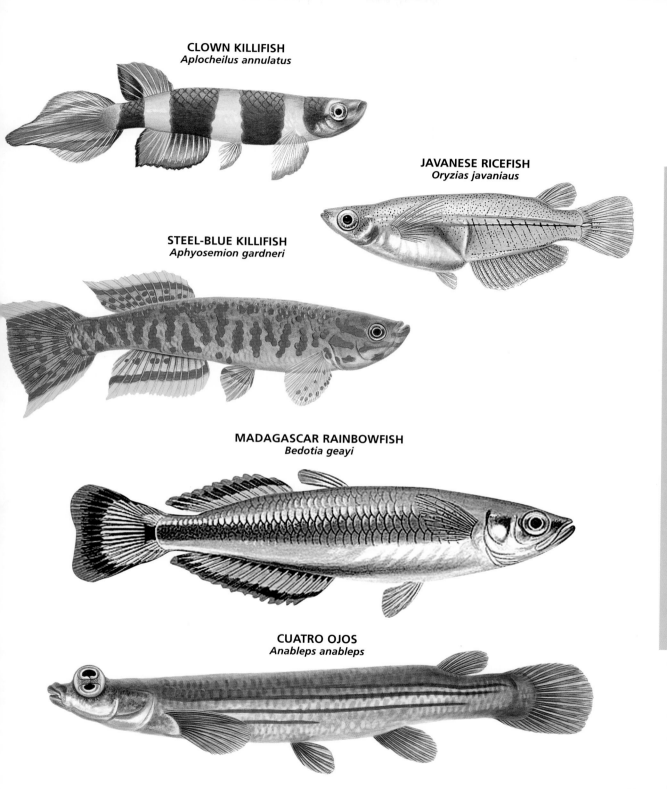

CLOWN KILLIFISH
Aplocheilus annulatus

JAVANESE RICEFISH
Oryzias javaniaus

STEEL-BLUE KILLIFISH
Aphyosemion gardneri

MADAGASCAR RAINBOWFISH
Bedotia geayi

CUATRO OJOS
Anableps anableps

Oarfishes, Squirrelfishes and Dories

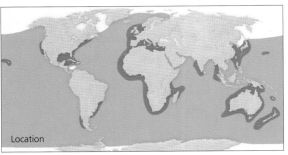

Location

- Oarfishes and squirrelfishes
- Oarfishes, squirrelfishes and dories

FLASHING ON AND OFF

Flashlight fishes have light organs which contain luminous bacteria. To hide from predators, the fish covers the light organ with a type of eyelid called a melanphore.

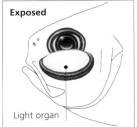

Exposed

Light organ

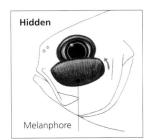

Hidden

Melanphore

FLASHLIGHT FISH
Photoblepharon palpebratus

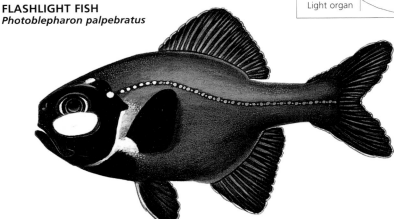

OARFISH
Regalecus glesne

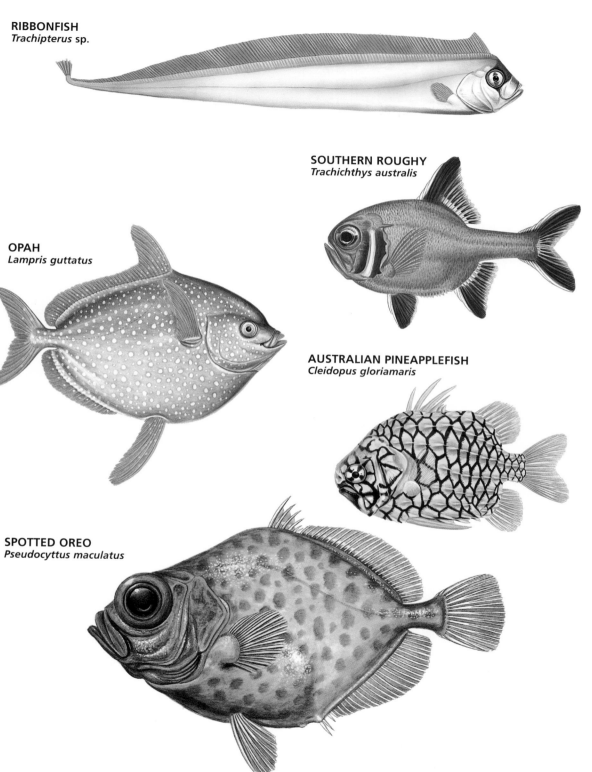

RIBBONFISH
Trachipterus sp.

SOUTHERN ROUGHY
Trachichthys australis

OPAH
Lampris guttatus

AUSTRALIAN PINEAPPLEFISH
Cleidopus gloriamaris

SPOTTED OREO
Pseudocyttus maculatus

477

Pipefishes, Swampeels and Scorpionfishes

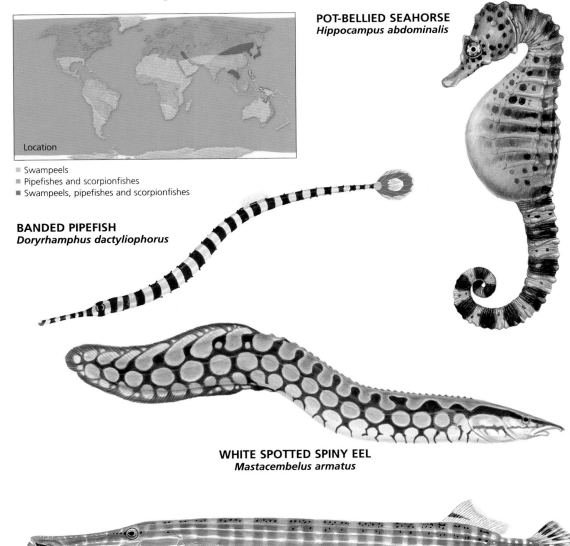

Location

- Swampeels
- Pipefishes and scorpionfishes
- Swampeels, pipefishes and scorpionfishes

POT-BELLIED SEAHORSE
Hippocampus abdominalis

BANDED PIPEFISH
Doryrhamphus dactyliophorus

WHITE SPOTTED SPINY EEL
Mastacembelus armatus

CARIBBEAN TRUMPETFISH
Aulostomus maculatus

SEAHORSE REPRODUCTION

The female lays eggs in the male's marsupial-like pouch, leaving them in his care. The eggs are incubated in his pouch until they hatch.

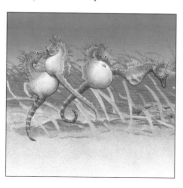

LUMPSUCKER
Cyclopterus lumpus

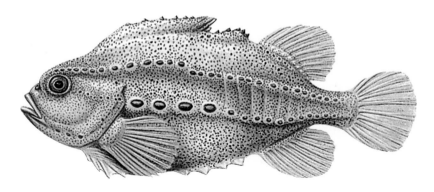

TIGER ROCKFISH
Sebastes nigrocinctus

COCKATOO WASPFISH
Ablabys taenionotus

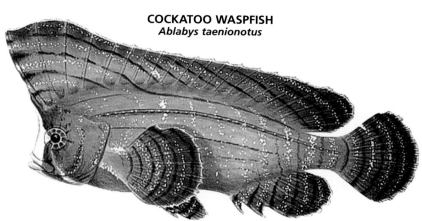

Perches, Groupers and Seabasses

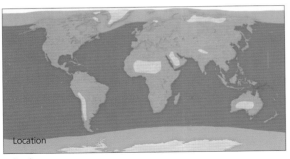

Location

■ Perches
■ Perches, groupers and seabasses

SPLENDID LICORICE GOURAMI
Parosphromenus dreissneri

SIXLINE SOAPFISH
Grammistes sexlineatus

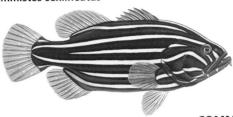

PURPLEQUEEN
Pseudanthis tuka

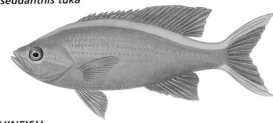

COMMON DOLPHINFISH
Coryphaena hippurus

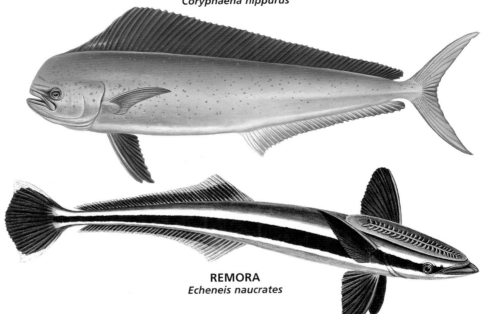

REMORA
Echeneis naucrates

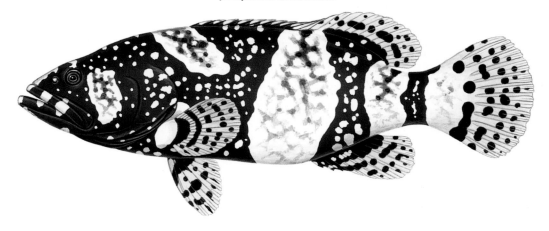

GIANT GROUPER
Epinephelus lanceolatus

SOUTH-EAST ASIA PIKEHEAD
Luciocephalus pulcher

MEYER'S BUTTERFLYFISH
Chaetodon meyeri

MOORISH IDOL
Zanclus cornutus

Cichlids, Damselfishes, Wrasses, Parrotfishes and Blennies

![map]

Location

- Cichlids, damselfishes, wrasses and parrotfishes
- Cichlids, damselfishes, wrasses, parrotfishes and blennies

MUDSKIPPERS
The male protects eggs laid by the female by wrapping his body around them.

CHANGING COLOR AND SEX
As they mature, highfin parrotfishes (*Scarus altipinnis*) travel through three color phases, which also indicate changes to gender.

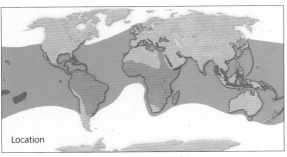

Juvenile: usually asexual female

Initial phase: usually female

Terminal phase: always a mature male

HARLEQUIN TUSKFISH
Choerodon fasciatus

STRIPED JULIE
Julidochromis regani

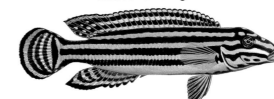

BLACK-HEADED BLENNY
Lipophrys nigriceps

OSCAR
Astronotus ocellatus

SPINECHEEK ANEMONEFISH
Premnas biaculeatus

RAINBOW CALE
Odax acroptilus

LAKE MALAWI ZEBRA CICHLID
Pseudotropheus zebra

Gobies, Flatfishes and Triggerfishes

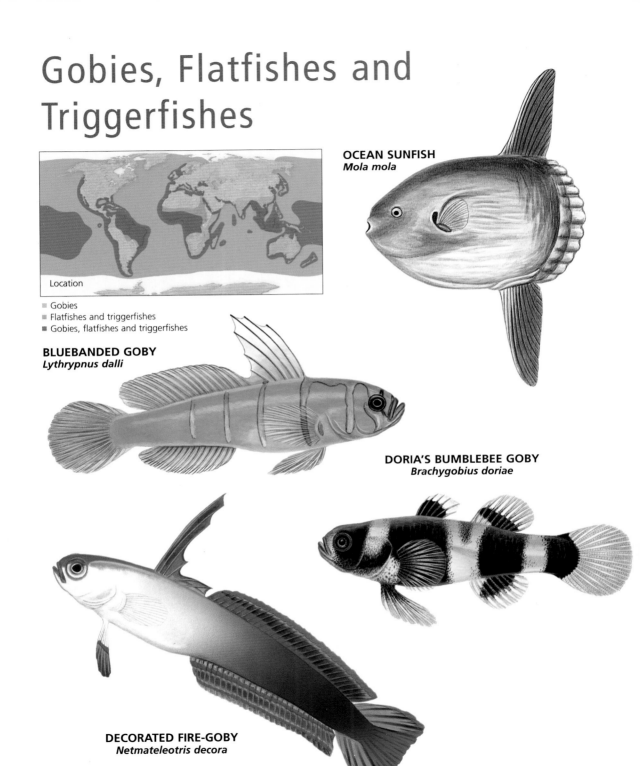

Location

- Gobies
- Flatfishes and triggerfishes
- Gobies, flatfishes and triggerfishes

OCEAN SUNFISH
Mola mola

BLUEBANDED GOBY
Lythrypnus dalli

DORIA'S BUMBLEBEE GOBY
Brachygobius doriae

DECORATED FIRE-GOBY
Netmateleotris decora

RIGHT-EYED FLATFISH

The left eye of right-eyed flatfishes (flounders) moves toward the right eye. The front of the skull twists to bring the jaws sideways.

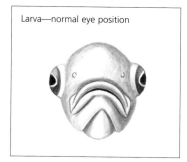

Larva—normal eye position

Left eye moves to top of head.

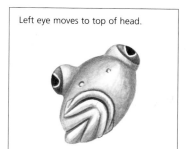

Adult—both eyes on right side

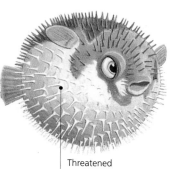

PEACOCK FLOUNDER
Bothus lunatus

PUFFERFISH

Pufferfishes inflate themselves into a spiny globe when threatened.

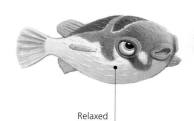

Relaxed

Threatened

SURVIVAL TECHNIQUES

The mimic filefish (*Paraluteres prionurus*) has evolved to look like the toxic blacksaddled puffer (*Canthigaster valentini*).

Blacksaddled puffer
(venomous)

Mimic filefish
(non-venomous)

FIGURE-EIGHT PUFFER
Tetraodon biocellatus

CLOWN TRIGGERFISH
Balistoides conspicillum

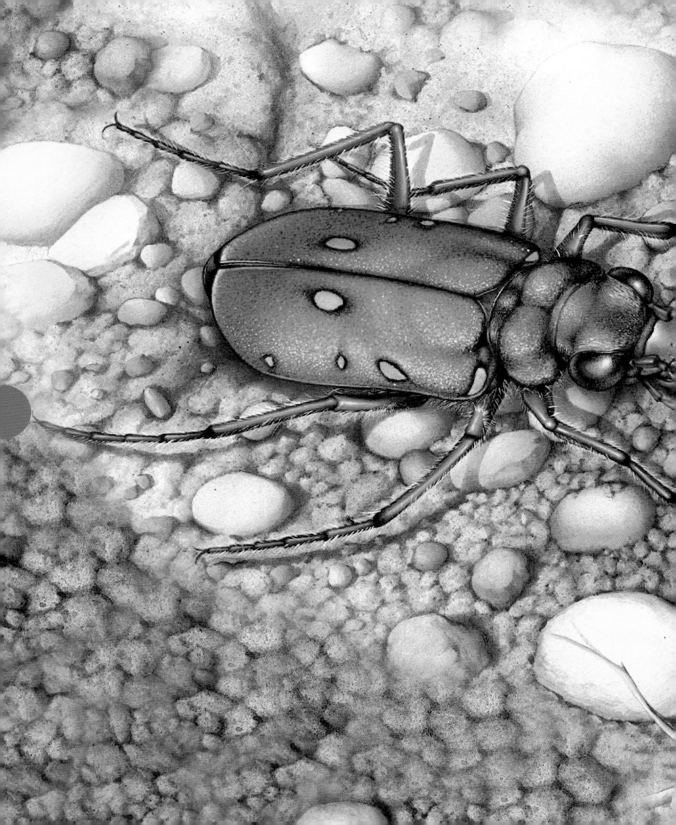

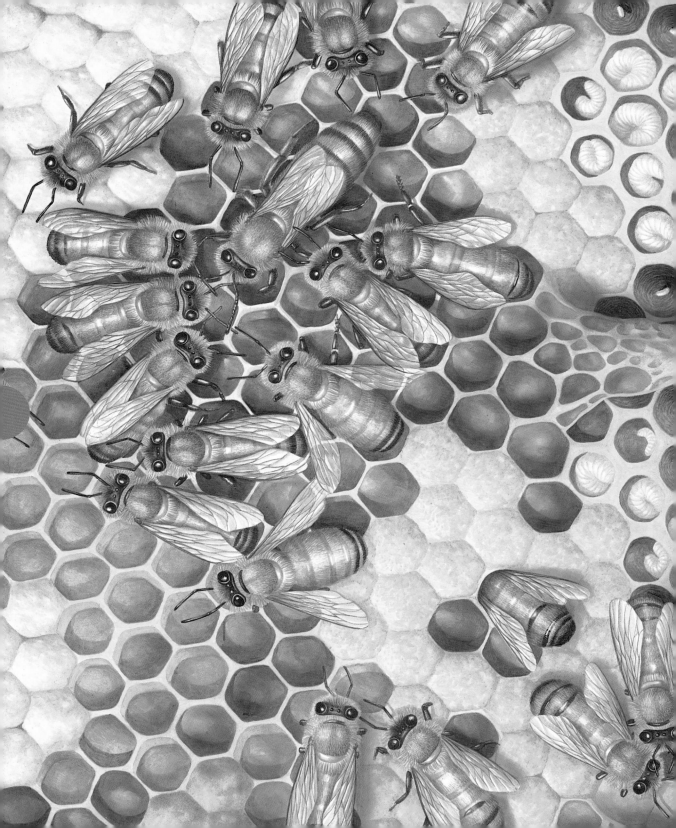

Classifying Invertebrates

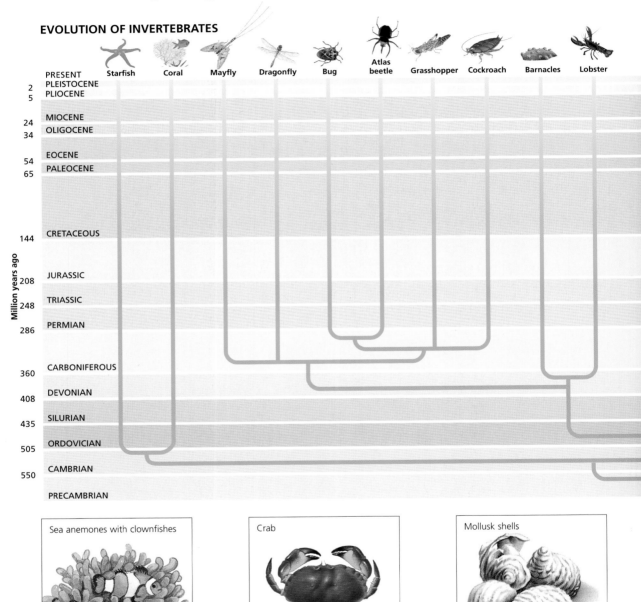

EVOLUTION OF INVERTEBRATES

| | Starfish | Coral | Mayfly | Dragonfly | Bug | Atlas beetle | Grasshopper | Cockroach | Barnacles | Lobster |

Million years ago	
PRESENT	
2	PLEISTOCENE
5	PLIOCENE
24	MIOCENE
34	OLIGOCENE
54	EOCENE
65	PALEOCENE
144	CRETACEOUS
208	JURASSIC
248	TRIASSIC
286	PERMIAN
360	CARBONIFEROUS
408	DEVONIAN
435	SILURIAN
505	ORDOVICIAN
550	CAMBRIAN
	PRECAMBRIAN

Sea anemones with clownfishes

Crab

Mollusk shells

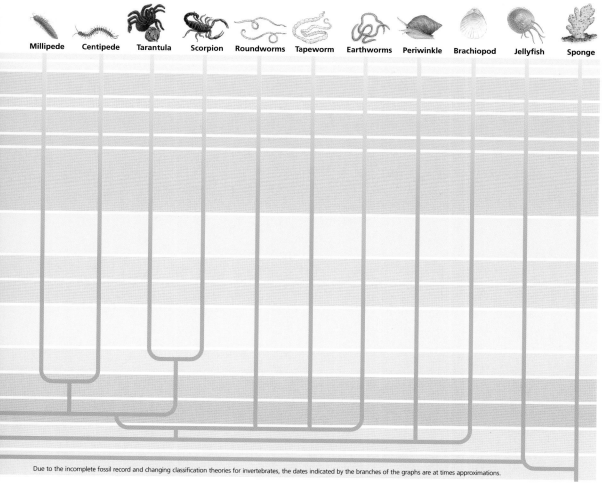

Millipede Centipede Tarantula Scorpion Roundworms Tapeworm Earthworms Periwinkle Brachiopod Jellyfish Sponge

Due to the incomplete fossil record and changing classification theories for invertebrates, the dates indicated by the branches of the graphs are at times approximations.

Weevil

Butterfly

Redback spider

Invertebrate Chordates, Sponges and Cnidarians

CORAL REEF ECOSYSTEM

Corals thrive in the shallow waters of the tropical Pacific and Indian oceans. Thousands of individual coral polyps, each protected by a limestone exoskeleton, group together to form living colonies; these in turn form coral reefs.

The crown-of-thorns starfish attacks and eats corals by disgorging its stomach over a colony, then absorbing the liquefied tissues.

Coral reefs provide shelter, food and breeding territory for thousands of species of plants and animals, such as fish, sharks and turtles.

Within certain coral communities, individual polyps have specialized functions—feeding, breeding or defense.

COLONIAL SEA SQUIRT
Didemnum molle

SPONGE
Spongia officinalis

SOFT CORAL
Lophelia pertusa

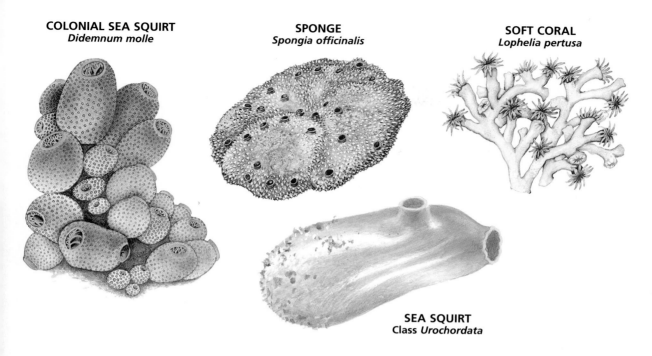

SEA SQUIRT
Class *Urochordata*

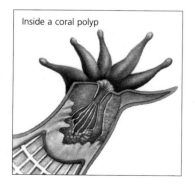

Sea anemone

Sea sponge

Portuguese man o'war

Inside a coral polyp

Jellyfish

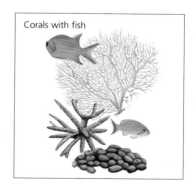

Corals with fish

493

Mollusks

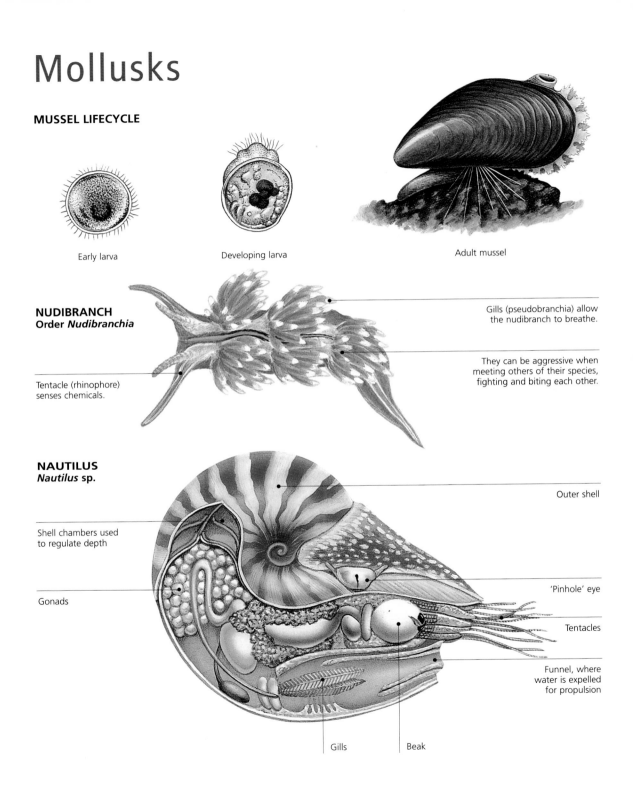

MUSSEL LIFECYCLE

Early larva

Developing larva

Adult mussel

NUDIBRANCH
Order *Nudibranchia*

Tentacle (rhinophore)
senses chemicals.

Gills (pseudobranchia) allow
the nudibranch to breathe.

They can be aggressive when
meeting others of their species,
fighting and biting each other.

NAUTILUS
Nautilus sp.

Shell chambers used
to regulate depth

Gonads

Outer shell

'Pinhole' eye

Tentacles

Funnel, where
water is expelled
for propulsion

Gills

Beak

Common garden snail

Periwinkle (marine snail)

Periwinkle displaying muscular foot

Squid

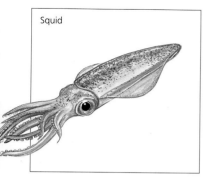

Oyster shells

Oyster with pearl—layers of mucus surrounding foreign body

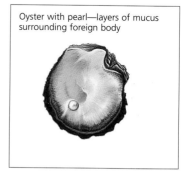

Brachiopod shells

Limpet displaying muscular foot

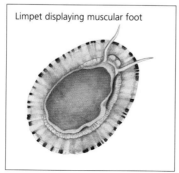

Mussels

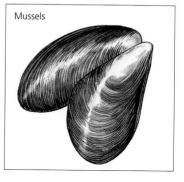

Bivalve shells

Mud whelk

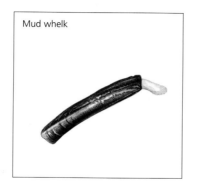

Common garden slug

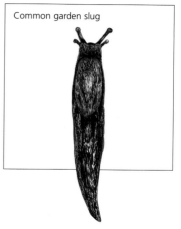

Squids and Octopuses

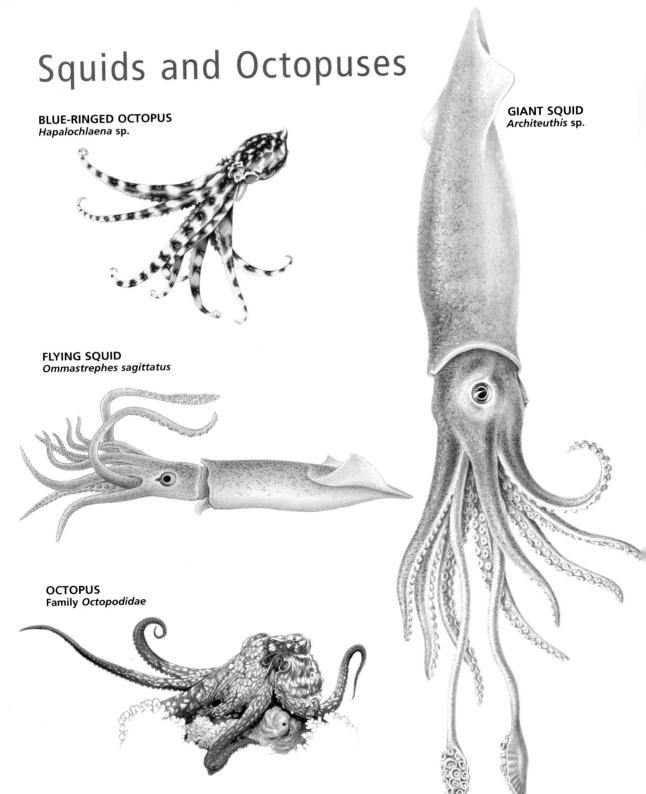

BLUE-RINGED OCTOPUS
Hapalochlaena sp.

GIANT SQUID
Architeuthis sp.

FLYING SQUID
Ommastrephes sagittatus

OCTOPUS
Family *Octopodidae*

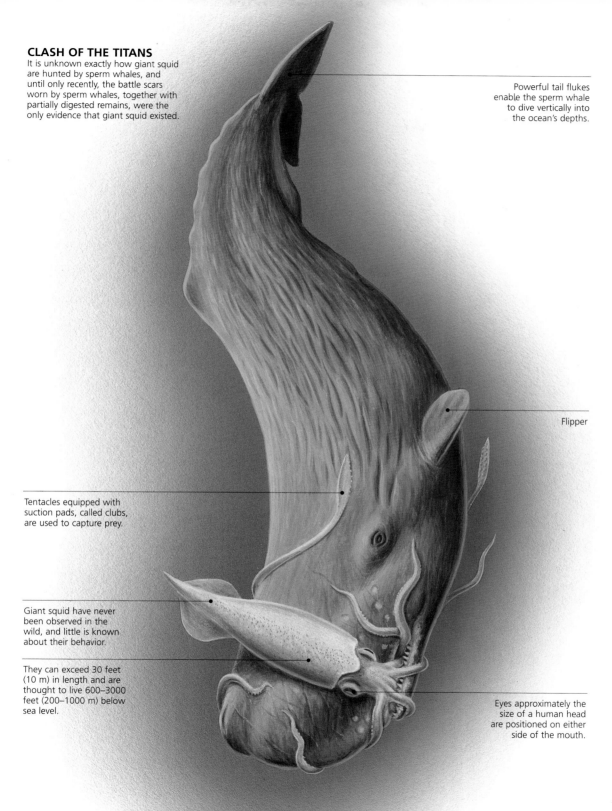

CLASH OF THE TITANS

It is unknown exactly how giant squid are hunted by sperm whales, and until only recently, the battle scars worn by sperm whales, together with partially digested remains, were the only evidence that giant squid existed.

Powerful tail flukes enable the sperm whale to dive vertically into the ocean's depths.

Flipper

Tentacles equipped with suction pads, called clubs, are used to capture prey.

Giant squid have never been observed in the wild, and little is known about their behavior.

They can exceed 30 feet (10 m) in length and are thought to live 600–3000 feet (200–1000 m) below sea level.

Eyes approximately the size of a human head are positioned on either side of the mouth.

Worms

PARASITIC WORMS

Hookworm eggs

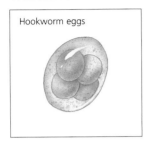

Tapeworm head

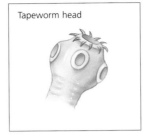

WHIPWORMS
Trichuris sp.

MARINE TAPEWORM
Class *Cestoidea*

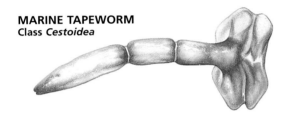

MARINE LEECH
Family *Piscicolidae*

ROUNDWORMS
Phylum *Nematoda*

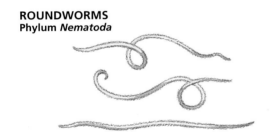

TERRESTRIAL TAPEWORM
Class *Cestoidea*

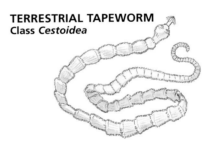

EARTHWORMS
Class *Oligochaeta*

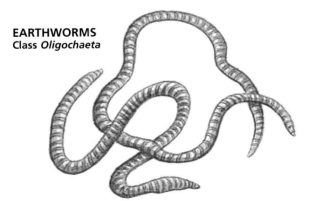

WORM EATING APPLE
While many worms are parasitic (they survive by living off their host), others eat plant or animal matter, such as fruit.

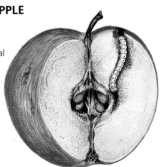

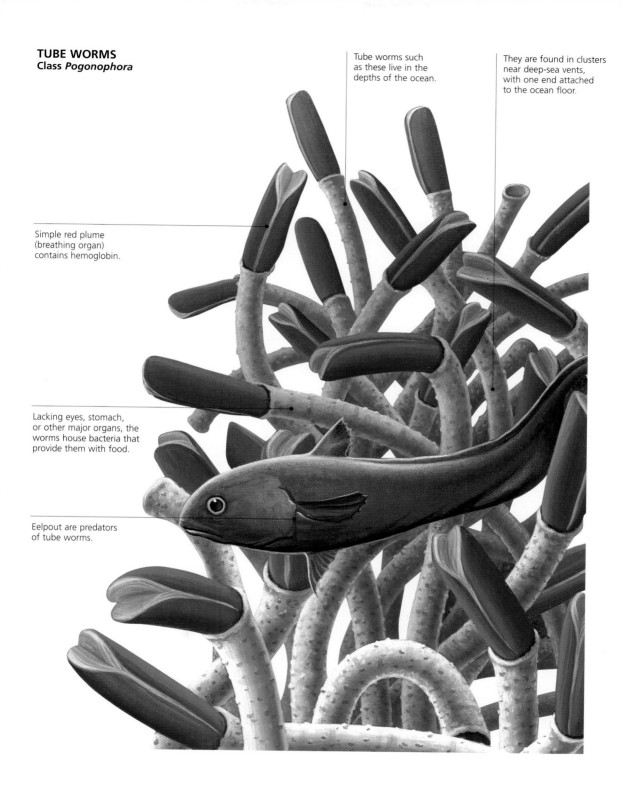

TUBE WORMS
Class *Pogonophora*

Tube worms such as these live in the depths of the ocean.

They are found in clusters near deep-sea vents, with one end attached to the ocean floor.

Simple red plume (breathing organ) contains hemoglobin.

Lacking eyes, stomach, or other major organs, the worms house bacteria that provide them with food.

Eelpout are predators of tube worms.

Echinoderms

STARFISH ANATOMY

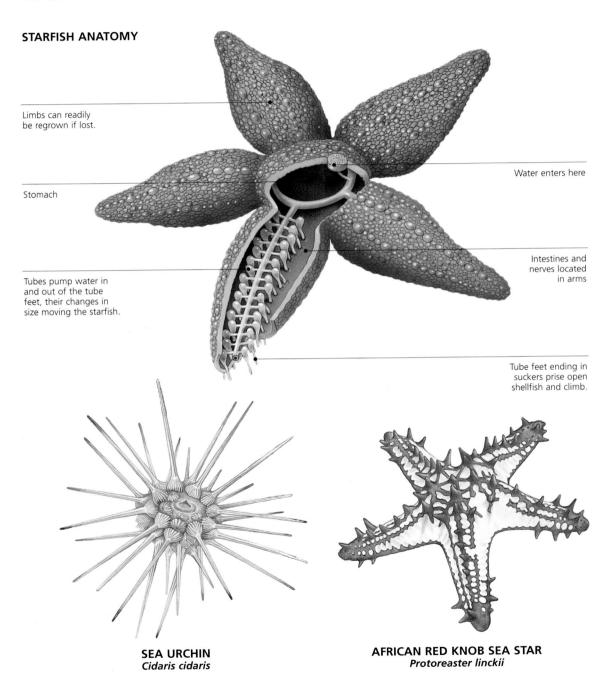

Limbs can readily
be regrown if lost.

Water enters here

Stomach

Tubes pump water in
and out of the tube
feet, their changes in
size moving the starfish.

Intestines and
nerves located
in arms

Tube feet ending in
suckers prise open
shellfish and climb.

SEA URCHIN
Cidaris cidaris

AFRICAN RED KNOB SEA STAR
Protoreaster linckii

SEA CUCUMBER
Stichopus chloronotus

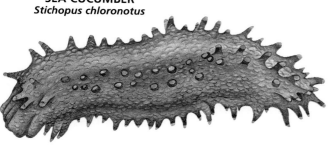

CROWN-OF-THORNS STARFISH
Acanthaster planci

Nocturnal crown-of-thorns starfish feed on coral polyps when corals are most active.

Individual starfish can devour over 10 square miles (26 sq km) annually.

Sea star

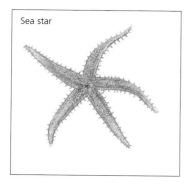

Starfish

Baculogypsina

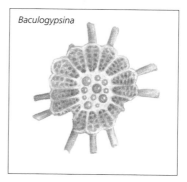

Calcarina

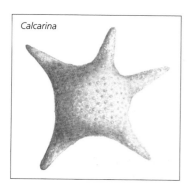

Starfish

Sea urchin

Arthropods

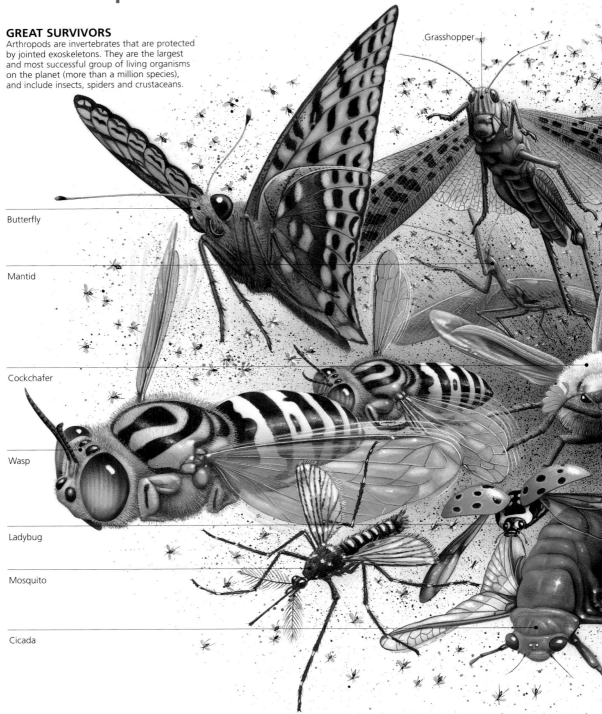

GREAT SURVIVORS

Arthropods are invertebrates that are protected by jointed exoskeletons. They are the largest and most successful group of living organisms on the planet (more than a million species), and include insects, spiders and crustaceans.

Grasshopper

Butterfly

Mantid

Cockchafer

Wasp

Ladybug

Mosquito

Cicada

Bug

Mayfly

Fly   Lacewing   Dragonfly

Spider

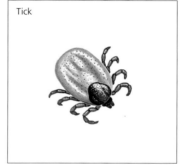

Tick

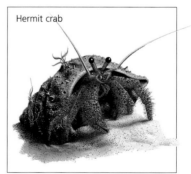

Hermit crab

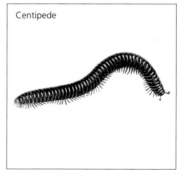

Centipede

503

Arachnids

EVOLUTION OF ARACHNIDS

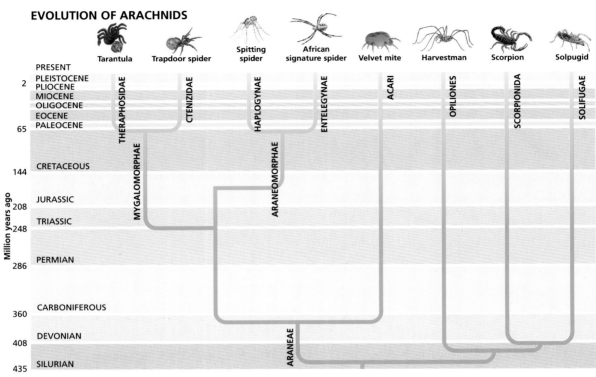

Tarantula	Trapdoor spider	Spitting spider	African signature spider	Velvet mite	Harvestman	Scorpion	Solpugid

PRESENT
PLEISTOCENE
PLIOCENE
MIOCENE
OLIGOCENE
EOCENE
PALEOCENE

2

65

CRETACEOUS 144

JURASSIC 208

TRIASSIC 248

PERMIAN 286

CARBONIFEROUS 360

DEVONIAN 408

SILURIAN 435

Million years ago

THERAPHOSIDAE
CTENIZIDAE
HAPLOGYNAE
ENTELEGYNAE
ACARI
OPILIONES
SCORPIONIDA
SOLIFUGAE
MYGALOMORPHAE
ARANEOMORPHAE
ARANEAE

ARACHNID ANATOMY

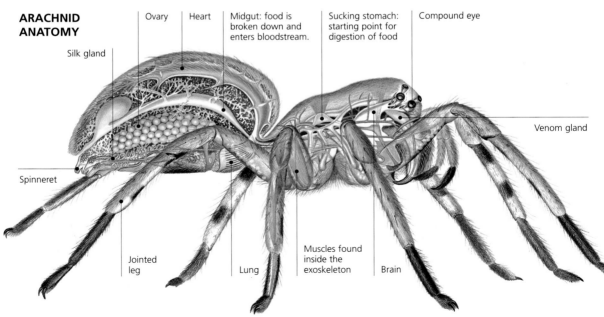

Silk gland

Ovary

Heart

Midgut: food is broken down and enters bloodstream.

Sucking stomach: starting point for digestion of food

Compound eye

Venom gland

Spinneret

Jointed leg

Lung

Muscles found inside the exoskeleton

Brain

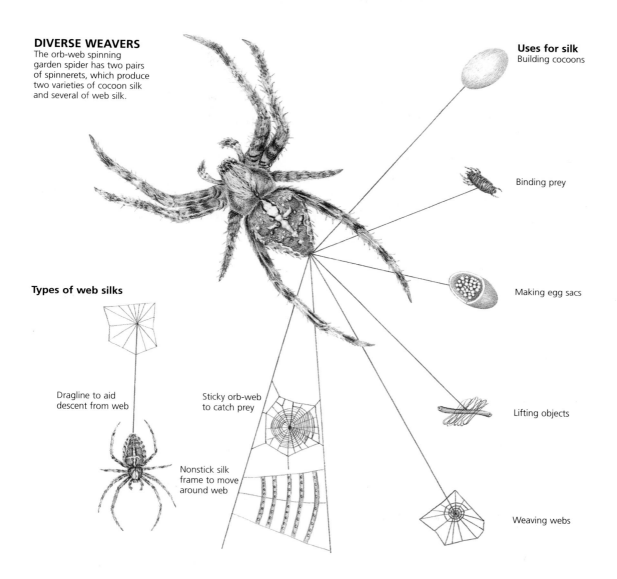

DIVERSE WEAVERS

The orb-web spinning garden spider has two pairs of spinnerets, which produce two varieties of cocoon silk and several of web silk.

Uses for silk

Building cocoons

Binding prey

Making egg sacs

Lifting objects

Weaving webs

Types of web silks

Dragline to aid descent from web

Sticky orb-web to catch prey

Nonstick silk frame to move around web

SPIDER FACES

Crab spider

Woodlouse-eating spider

Ogre-faced spider

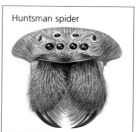

Huntsman spider

Spiders

Arthropods

INVERTEBRATES

WHITE LADY SPIDER
Family *Sparassidae*

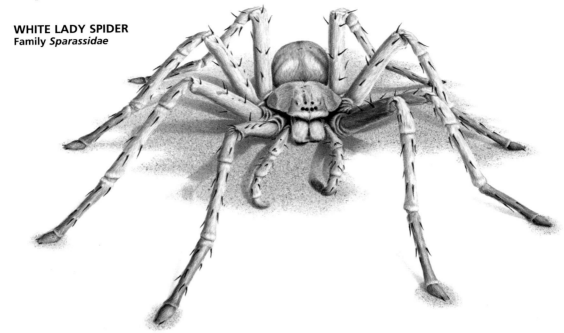

PRIMITIVE SPIDER
Suborder *Labidognatha*

COMB-FOOTED SPIDER
Family *Theridiidae*

JUMPING SPIDER
Family *Salticidae*

TARANTULA
Family *Theraphosidae*

LYNX SPIDER
Family *Oxyopidae*

CRAB SPIDER
Family *Thomisidae*

COURTSHIP DISPLAYS

Wolf spiders

Jumping spiders

REPRODUCTION RITUALS

Nursery web spiders
Male presents wrapped prey
to female, before mating.

Crab spiders
Male fastens female to ground
with silk, then deposits sperm.

SPIDER JAWS

Mygalomorphic (hairy) spider
Fangs point downward

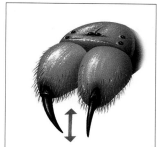

Arabeomorphic (typical) spider
Fangs hinge together sideways

SHEDDING THE EXOSKELETON

Hangs from web

Exoskeleton tears apart

Draws fragile legs out

Waits for exoskeleton to dry

Spiders

REDBACK SPIDER
Family *Theridiidae*

BOLAS SPIDER
Family *Araneidae*

NORTHERN BLACK WIDOW
Family *Theridiidae*

SPITTING SPIDER
Family *Scytodidae*

JUMPING SPIDERS
Family *Salticidae*

MOUSE SPIDER
Family *Theraphosidae*

TRAPDOOR SPIDER
Family *Ctenizidae*

FUNNEL-WEB SPIDER
Family *Agelenidae*

VIOLIN SPIDER
Family *Loxoscelidae*

FROM EGG TO ADULT

Huntsman egg sac with eggs

Huntsman spiderling

Juvenile huntsman

Adult huntsman

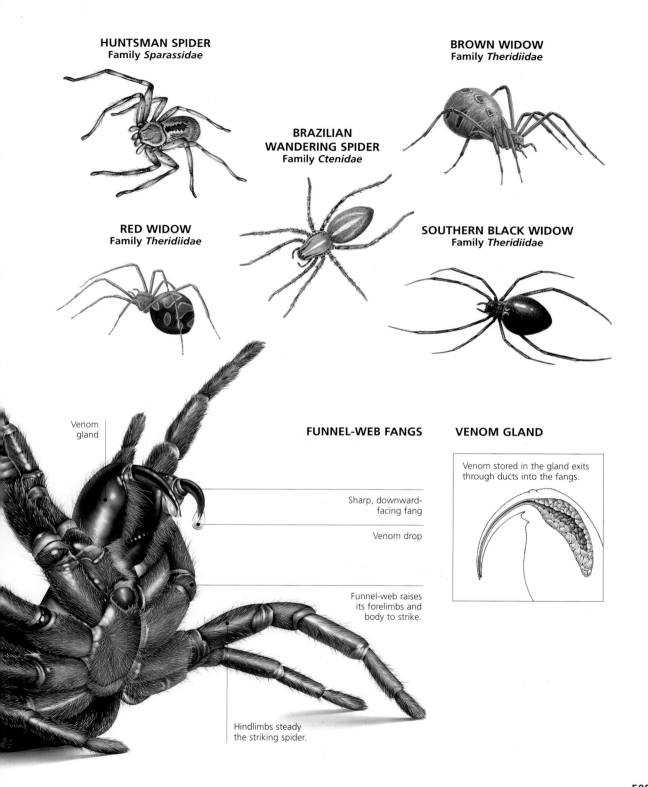

HUNTSMAN SPIDER
Family *Sparassidae*

BROWN WIDOW
Family *Theridiidae*

**BRAZILIAN
WANDERING SPIDER**
Family *Ctenidae*

RED WIDOW
Family *Theridiidae*

SOUTHERN BLACK WIDOW
Family *Theridiidae*

Venom
gland

FUNNEL-WEB FANGS

VENOM GLAND

Sharp, downward-
facing fang

Venom drop

Funnel-web raises
its forelimbs and
body to strike.

Venom stored in the gland exits
through ducts into the fangs.

Hindlimbs steady
the striking spider.

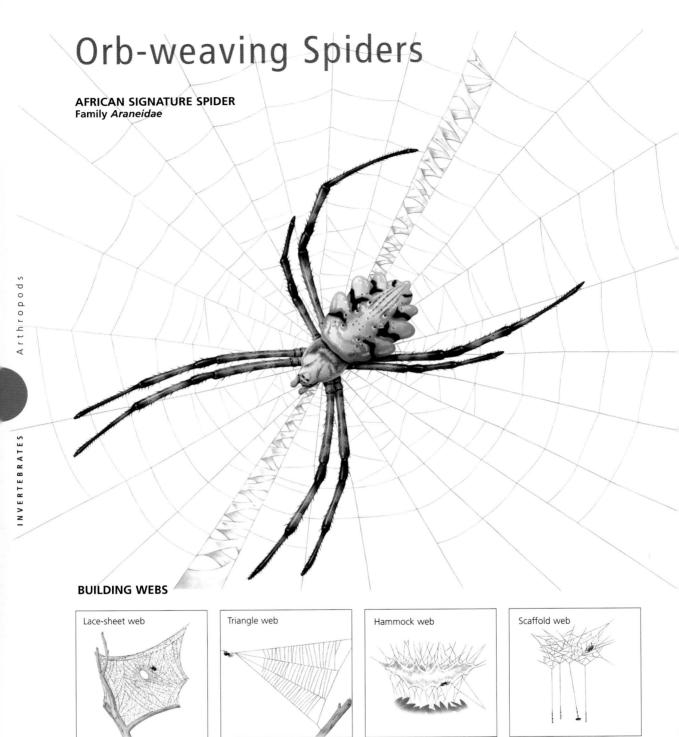

Orb-weaving Spiders

AFRICAN SIGNATURE SPIDER
Family *Araneidae*

BUILDING WEBS

Lace-sheet web

Triangle web

Hammock web

Scaffold web

MARBLED ORB WEAVER
Family *Araneidae*

SIGNATURE SPIDERS
Family *Araneidae*

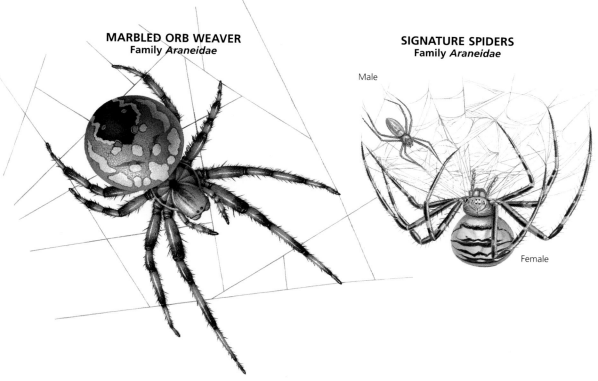

Male

Female

GOLDEN SILK SPIDER
Nephila sp.

SPINY ORB WEAVER
Family *Araneidae*

LONG-JAWED ORB WEAVER
Family *Araneidae*

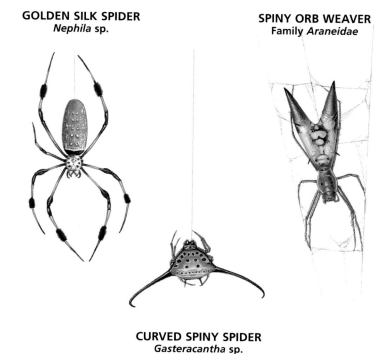

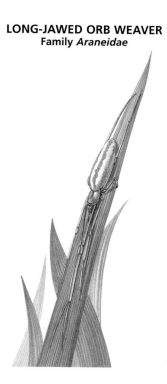

CURVED SPINY SPIDER
Gasteracantha sp.

Scorpions, Mites and Ticks

Arthropods

INVERTEBRATES

HARVESTMAN
Family *Leiobunidae*

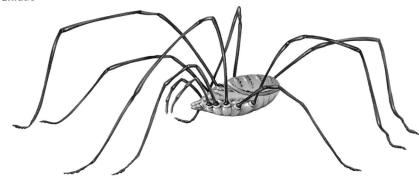

SOLPUGID
Family *Solpugidae*

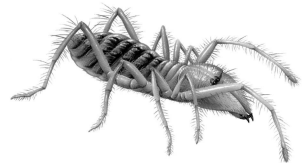

HARD TICK
Family *Ixodidae*

SCORPION
Family *Buthidae*

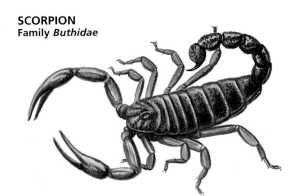

VELVET MITE
Family *Trombidiidae*

SCORPION ANATOMY

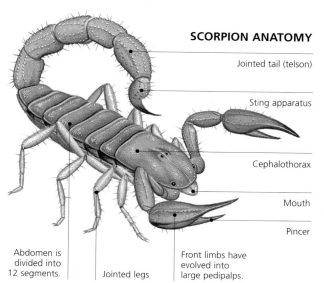

Jointed tail (telson)

Sting apparatus

Cephalothorax

Mouth

Pincer

Abdomen is divided into 12 segments.

Jointed legs

Front limbs have evolved into large pedipalps.

STING IN THE TAIL

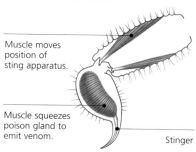

Muscle moves position of sting apparatus.

Muscle squeezes poison gland to emit venom.

Stinger

SCORPION COURTSHIP DISPLAY

Some species of scorpions perform elaborate courtship displays, which involve joining pincers and 'dancing' around each other before mating.

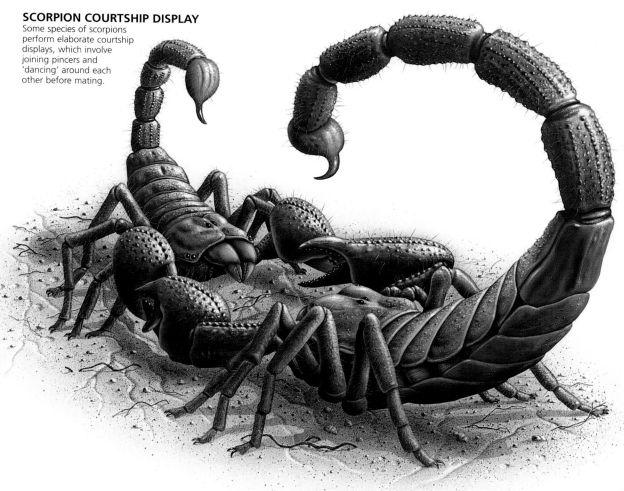

Crustaceans, Silverfishes, Centipedes and Millipedes

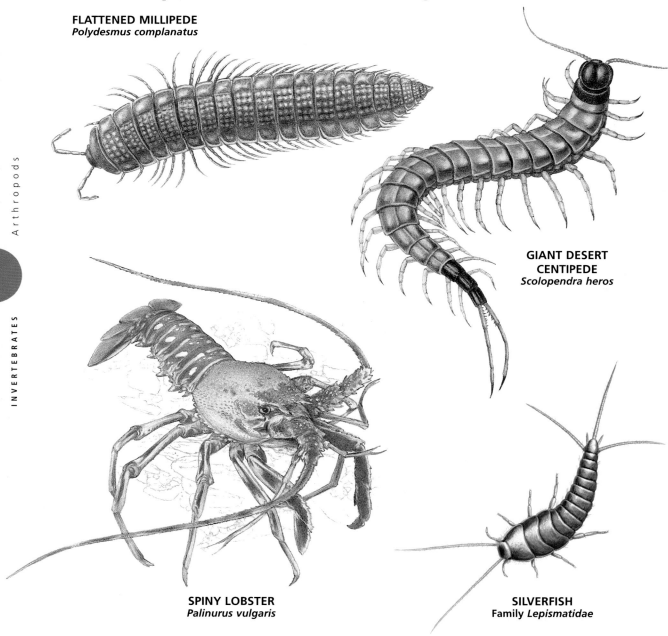

FLATTENED MILLIPEDE
Polydesmus complanatus

**GIANT DESERT
CENTIPEDE**
Scolopendra heros

SPINY LOBSTER
Palinurus vulgaris

SILVERFISH
Family *Lepismatidae*

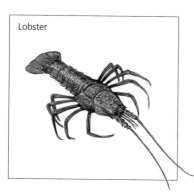
Lobster

Shrimp

Centipede

Crab

SPINY LOBSTER MIGRATION

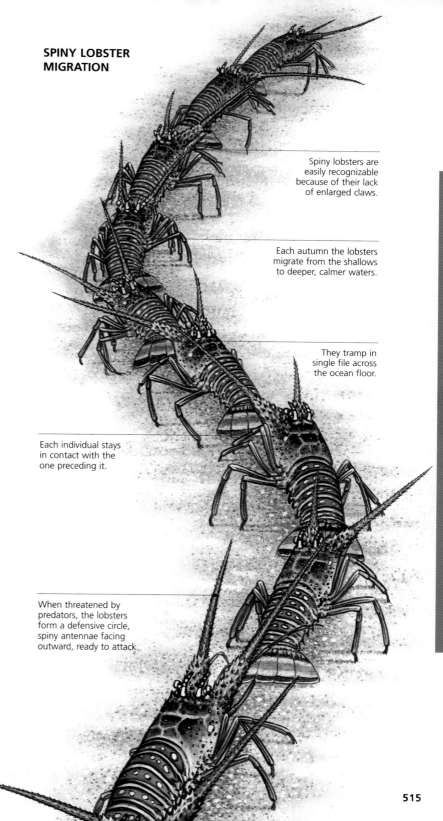

Spiny lobsters are easily recognizable because of their lack of enlarged claws.

Each autumn the lobsters migrate from the shallows to deeper, calmer waters.

They tramp in single file across the ocean floor.

Each individual stays in contact with the one preceding it.

When threatened by predators, the lobsters form a defensive circle, spiny antennae facing outward, ready to attack.

Crustaceans, Silverfishes, Centipedes and Millipedes

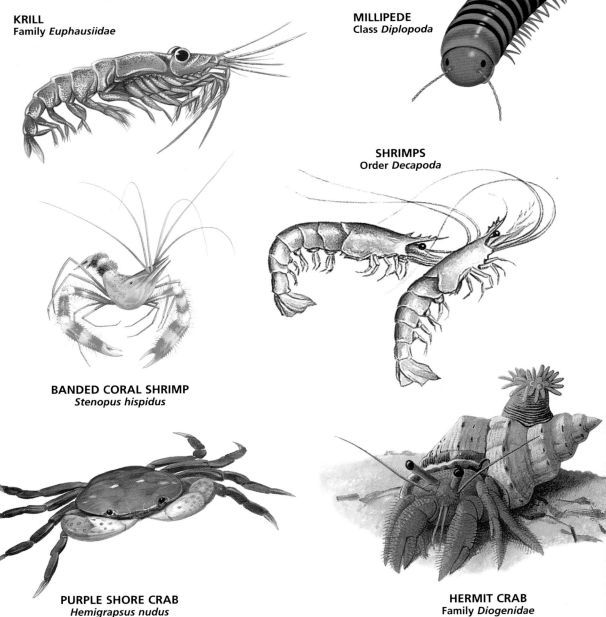

KRILL
Family *Euphausiidae*

MILLIPEDE
Class *Diplopoda*

SHRIMPS
Order *Decapoda*

BANDED CORAL SHRIMP
Stenopus hispidus

PURPLE SHORE CRAB
Hemigrapsus nudus

HERMIT CRAB
Family *Diogenidae*

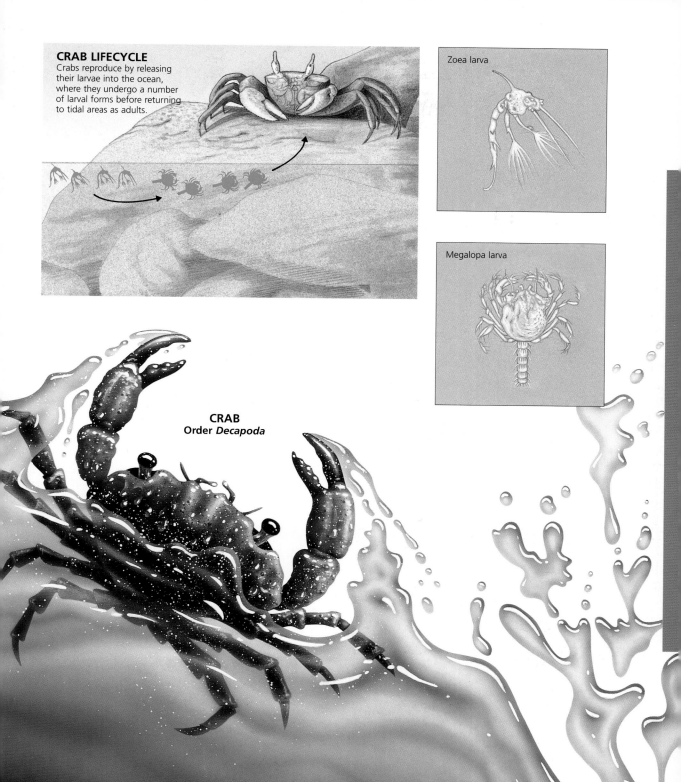

CRAB LIFECYCLE
Crabs reproduce by releasing their larvae into the ocean, where they undergo a number of larval forms before returning to tidal areas as adults.

Zoea larva

Megalopa larva

CRAB
Order *Decapoda*

Insects

EVOLUTION OF INSECTS

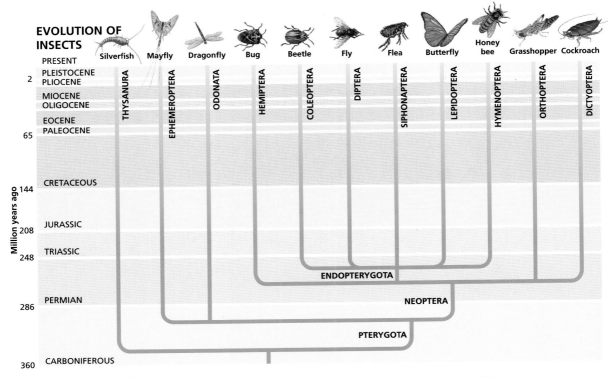

| | Silverfish | Mayfly | Dragonfly | Bug | Beetle | Fly | Flea | Butterfly | Honey bee | Grasshopper | Cockroach |

PRESENT
PLEISTOCENE
PLIOCENE
2
MIOCENE
OLIGOCENE
EOCENE
PALEOCENE
65
CRETACEOUS 144
JURASSIC 208
TRIASSIC 248
PERMIAN 286
CARBONIFEROUS 360

Million years ago

THYSANURA · EPHEMEROPTERA · ODONATA · HEMIPTERA · COLEOPTERA · DIPTERA · SIPHONAPTERA · LEPIDOPTERA · HYMENOPTERA · ORTHOPTERA · DICTYOPTERA

ENDOPTERYGOTA

NEOPTERA

PTERYGOTA

TYPES OF WINGS

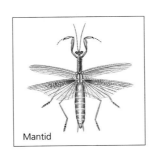

True fly

Mantid

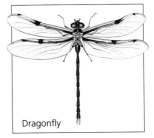

Butterfly

Thrip

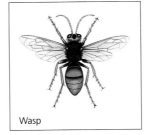

Wasp

Dragonfly

INSECT ANATOMY

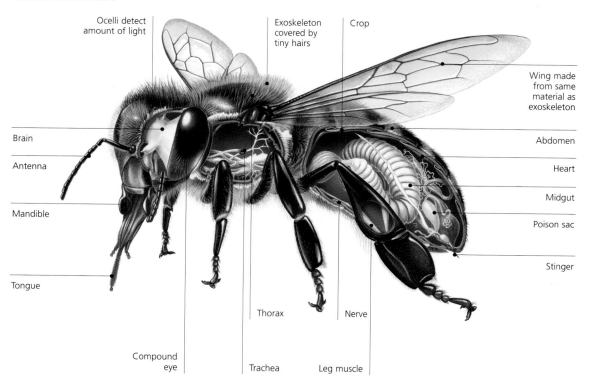

Ocelli detect amount of light

Exoskeleton covered by tiny hairs

Crop

Wing made from same material as exoskeleton

Brain

Antenna

Mandible

Abdomen

Heart

Midgut

Poison sac

Stinger

Tongue

Thorax

Nerve

Compound eye

Trachea

Leg muscle

TYPES OF ANTENNAE

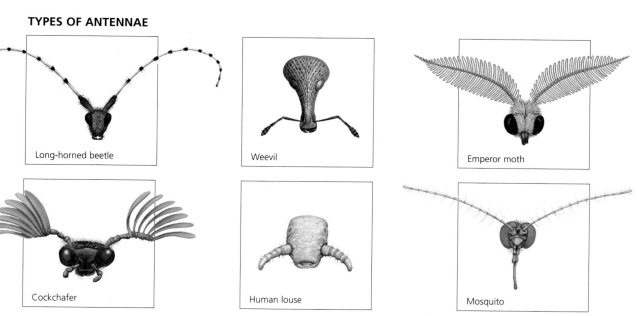

Long-horned beetle

Weevil

Emperor moth

Cockchafer

Human louse

Mosquito

Dragonflies, Mayflies and Mantids

Arthropods

INVERTEBRATES

DRAGONFLY ANATOMY

Ocelli monitor the horizon to keep dragonfly flying level.

Front wing

Rear wing beats at a different speed to front wing to increase stability.

Each lens of the compound eye sees things differently.

Antenna

Jointed leg

Hairs around mouth taste and guide food.

DRAGONFLY LIFECYCLE

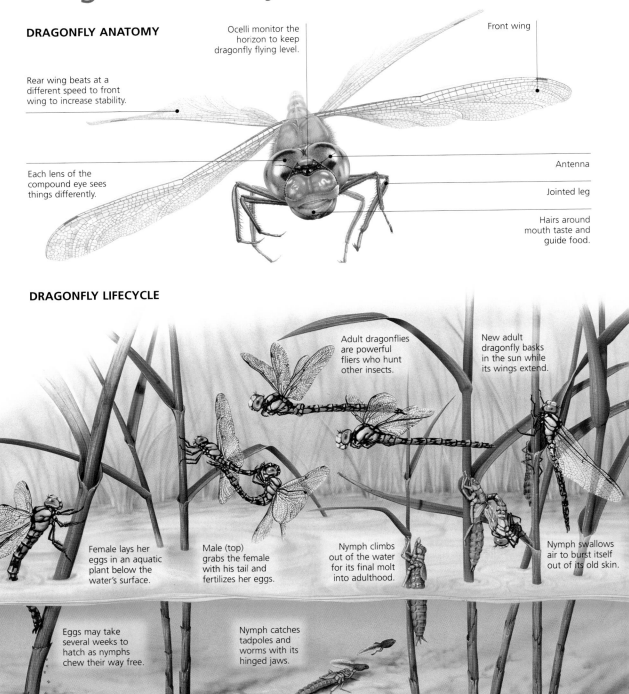

Adult dragonflies are powerful fliers who hunt other insects.

New adult dragonfly basks in the sun while its wings extend.

Female lays her eggs in an aquatic plant below the water's surface.

Male (top) grabs the female with his tail and fertilizes her eggs.

Nymph climbs out of the water for its final molt into adulthood.

Nymph swallows air to burst itself out of its old skin.

Eggs may take several weeks to hatch as nymphs chew their way free.

Nymph catches tadpoles and worms with its hinged jaws.

Praying mantis forelimb

Dragonfly

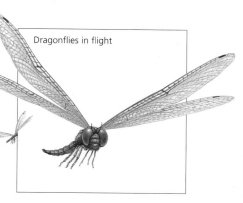

Dragonflies in flight

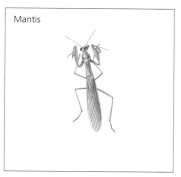

Mantis

Damselfly nymph

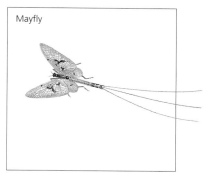

Mayfly

PRAYING MANTIS
Family *Mantidae*

ORCHID MANTIS
Family *Hymenopodidae*

Cockroaches, Termites and Lice

DAMP-WOOD TERMITE
Family
Rhinotermitidae

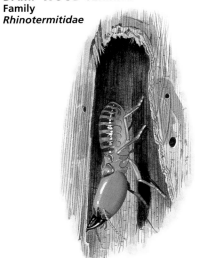

**FEMALE COCKROACH
LAYING EGGS**
As the eggs of many species
of cockroaches are laid on
the ground, they are covered
in hard cases for protection.

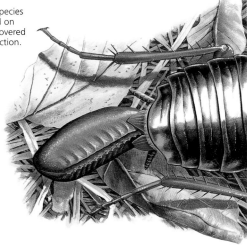

Cockroach

Termite

Spinifex termite

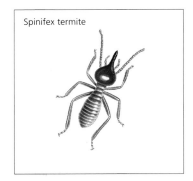

Cockroach head displaying antennae

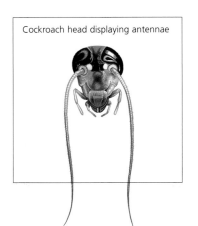

Louse

Fish louse

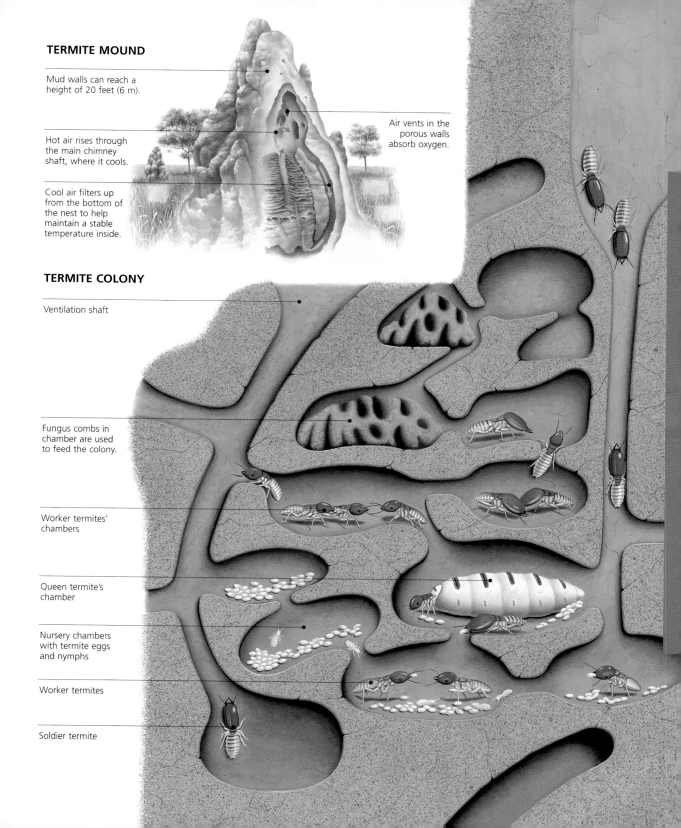

TERMITE MOUND

Mud walls can reach a height of 20 feet (6 m).

Air vents in the porous walls absorb oxygen.

Hot air rises through the main chimney shaft, where it cools.

Cool air filters up from the bottom of the nest to help maintain a stable temperature inside.

TERMITE COLONY

Ventilation shaft

Fungus combs in chamber are used to feed the colony.

Worker termites' chambers

Queen termite's chamber

Nursery chambers with termite eggs and nymphs

Worker termites

Soldier termite

Crickets and Grasshoppers

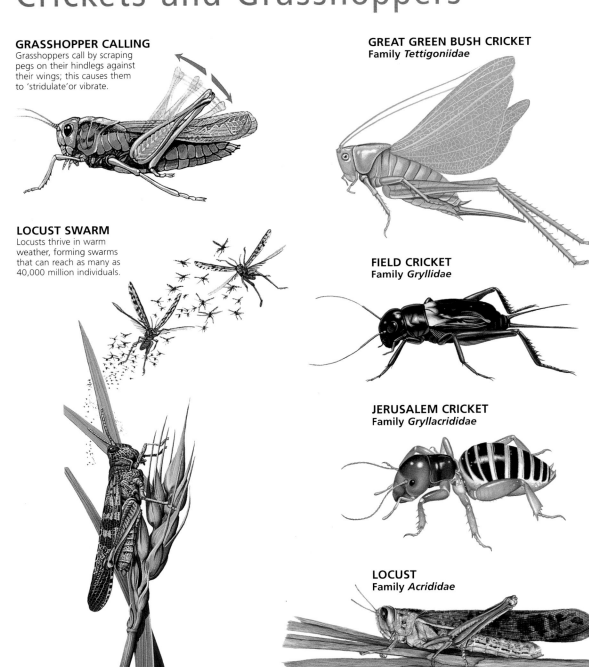

GRASSHOPPER CALLING
Grasshoppers call by scraping pegs on their hindlegs against their wings; this causes them to 'stridulate'or vibrate.

LOCUST SWARM
Locusts thrive in warm weather, forming swarms that can reach as many as 40,000 million individuals.

GREAT GREEN BUSH CRICKET
Family *Tettigoniidae*

FIELD CRICKET
Family *Gryllidae*

JERUSALEM CRICKET
Family *Gryllacrididae*

LOCUST
Family *Acrididae*

Male mole cricket in tunnel

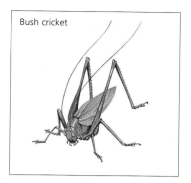
Bush cricket

Koringkriek cricket leg

True katydid on leaf

Grasshopper spreading wings

Grasshopper

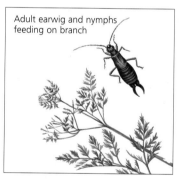
Adult earwig and nymphs feeding on branch

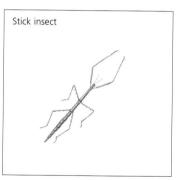
Stick insect

Grasshopper head

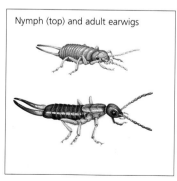
Nymph (top) and adult earwigs

Grasshopper leg

Nymph (top) and adult locusts

Bugs, Lacewings and Thrips

CICADA
Family *Cicadidae*

Anatomy of feeding
Bugs have distinctive mouthparts—a beaklike rostrum surrounded by four sharp stylets. The stylets pierce the food source, then the rostrum carries toxic saliva through the puncture to partially digest the meal before ingesting.

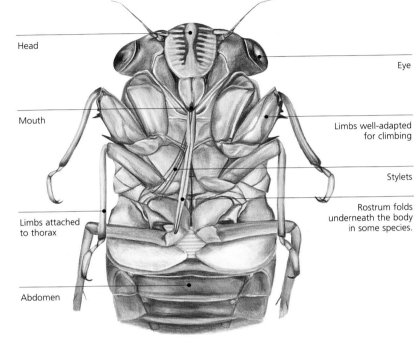

Head

Eye

Mouth

Limbs well-adapted for climbing

Stylets

Rostrum folds underneath the body in some species.

Limbs attached to thorax

Abdomen

LEAFHOPPER LIFECYCLE

Leafhopper waits on blade of grass for exoskeleton to harden.

Leafhopper nymph prepares to molt its final juvenile exoskeleton.

Newly adult leafhopper pulls itself free of its old exoskeleton.

Lacewing at rest

Lacewing taking off

Lacewing in flight

Stink bug
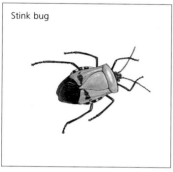

True bug with bright warning coloring

Milkweed bug

Lacewing

Thrip

Water bug leg

Nymph (top) and adult elder bugs
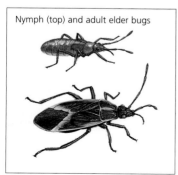

Periodical cicada
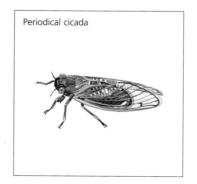

Assassin bug head
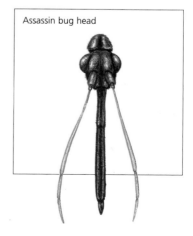

Bugs, Lacewings and Thrips

Arthropods

INVERTEBRATES

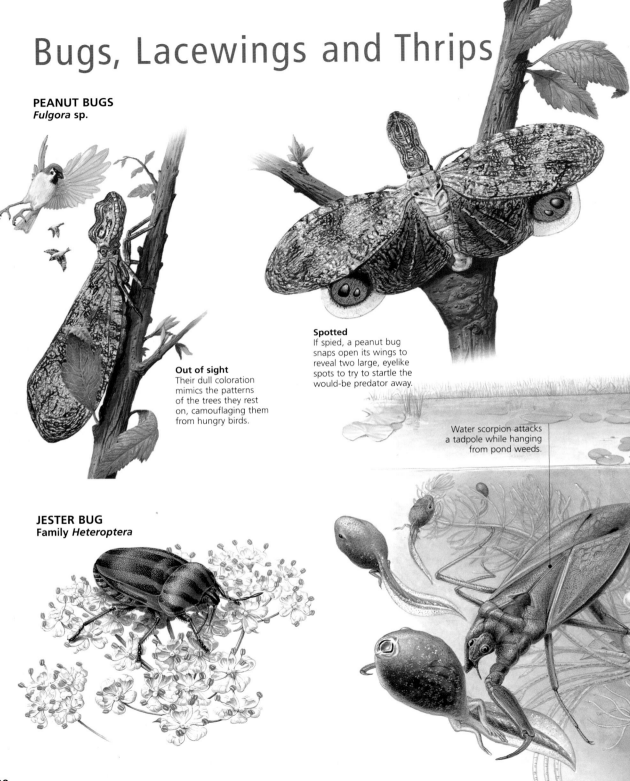

PEANUT BUGS
Fulgora sp.

Out of sight
Their dull coloration mimics the patterns of the trees they rest on, camouflaging them from hungry birds.

Spotted
If spied, a peanut bug snaps open its wings to reveal two large, eyelike spots to try to startle the would-be predator away.

Water scorpion attacks a tadpole while hanging from pond weeds.

JESTER BUG
Family *Heteroptera*

IN AND OUT OF WATER

Water boatman swims through the pond, using its legs as paddles.

Water striders use surface tension to 'walk' on the water's surface.

Many species of bugs spend larval or juvenile stages in the water.

Beetles

HERCULES BEETLES
Family *Dynastinae*
Rival male Hercules beetles grapple with their swordlike horns, battling for the right to mate with the female (right).

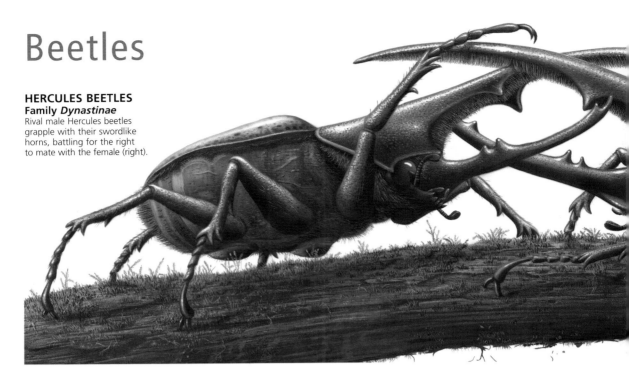

Diving beetle leg

Whirligig beetle swimming

Fire beetle

Diving beetle underwater

Darkling beetle

Jewel beetle

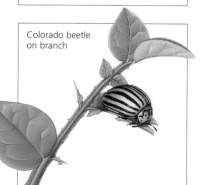

Colorado beetle

Flea beetle

Hercules beetle

Colorado beetle
on branch

Large chafer

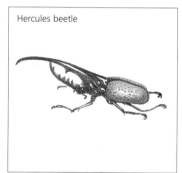

Rhinoceros beetle

Beetles

PAPUA NEW GUINEAN WEEVIL
Eupholus bennetti

TORTOISE BEETLE
Family *Chrysomelidae*

GIRAFFE WEEVIL
Family *Brentidae*

HARLEQUIN BEETLE
Family *Cerambycidae*

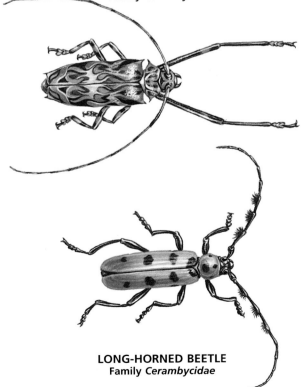

LONG-HORNED BEETLE
Family *Cerambycidae*

Female   **FIREFLIES**
Family *Lampyridae*   Male

BOMBARDIER BEETLE
Family *Carabidae*

DIVING BEETLE
Family *Dytiscidae*

VIOLIN BEETLE
Family *Buprestidae*

ROVE BEETLE
Family *Staphylinidae*

Ladybugs

TEN-SPOTTED LADYBUG
Family *Coccinellidae*

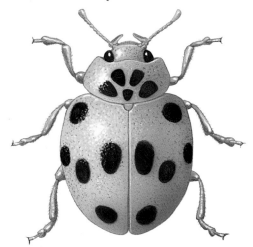

FIVE-SPOTTED LADYBUG
Family *Coccinellidae*

LADYBUG LAYING EGGS
The bright yellow eggs of the ladybug are laid on a leaf in clutches. The female chooses to lay her eggs close to a food source, such as an aphid colony.

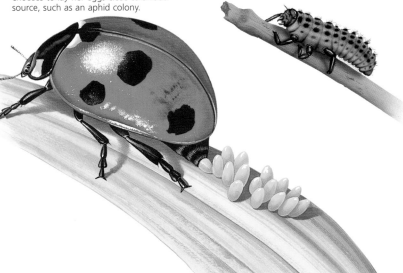

SEVEN-SPOTTED LADYBUG LARVA
Family *Coccinellidae*

LADYBUG LIFE STAGES

Egg
Hatches at 3–5 days

Larva
Pupates at 2–3 weeks

Pupa
Changes after 7–10 days

Adult
Lives for more than one year

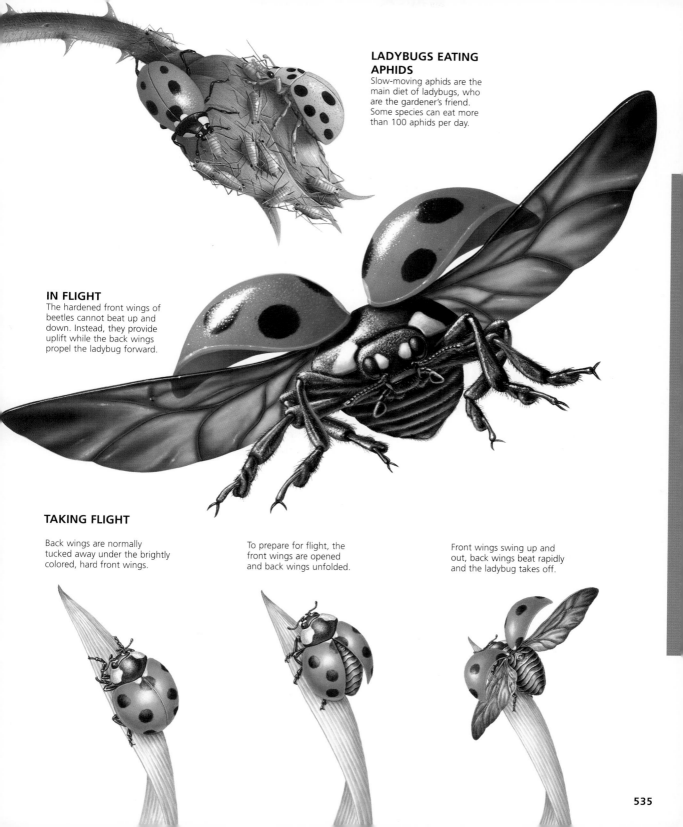

LADYBUGS EATING APHIDS

Slow-moving aphids are the main diet of ladybugs, who are the gardener's friend. Some species can eat more than 100 aphids per day.

IN FLIGHT

The hardened front wings of beetles cannot beat up and down. Instead, they provide uplift while the back wings propel the ladybug forward.

TAKING FLIGHT

Back wings are normally tucked away under the brightly colored, hard front wings.

To prepare for flight, the front wings are opened and back wings unfolded.

Front wings swing up and out, back wings beat rapidly and the ladybug takes off.

Flies, Fleas and Mosquitoes

DEER FLY
Family *Tabanidae*

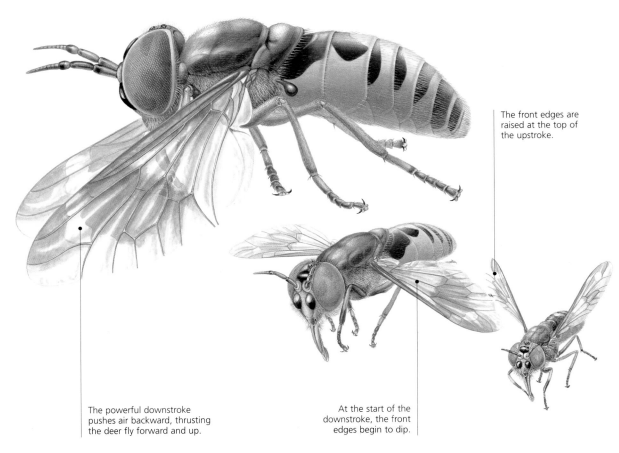

The front edges are
raised at the top of
the upstroke.

The powerful downstroke
pushes air backward, thrusting
the deer fly forward and up.

At the start of the
downstroke, the front
edges begin to dip.

CADDIS FLY LIFE STAGES

Egg	Larva	Pupa	Adult

Fly leg with sticky foot pads

Horsefly

Fly

Green bottle fly

House fly head

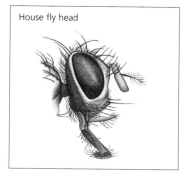

Hover fly

Blowfly maggots

Mosquito head

Midge

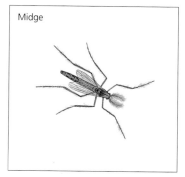

Horsefly head

Caddis fly

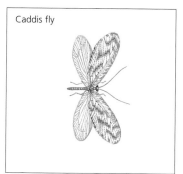

Flea jumping

537

Flies, Fleas and Mosquitoes

Arthropods

INVERTEBRATES

ANOPHELES MOSQUITO
Family *Culicidae*

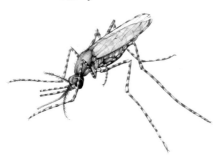

COMMON FLEA
Family *Pulicidae*

HOUSE FLY
Family *Muscidae*

Mosquito head: long proboscis is inserted into prey to suck blood.

Flea head: backward facing cheek combs identify species.

Fly head: colorfully patterned compound eyes aid navigation.

MOSQUITO LIFE STAGES

Culex laying eggs

Egg raft

Anopheles nymph

Culex nymph

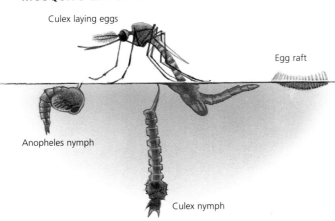

CULEX MOSQUITO
Culex sp.

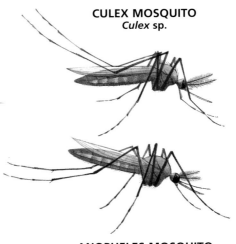

ANOPHELES MOSQUITO
Anopheles sp.

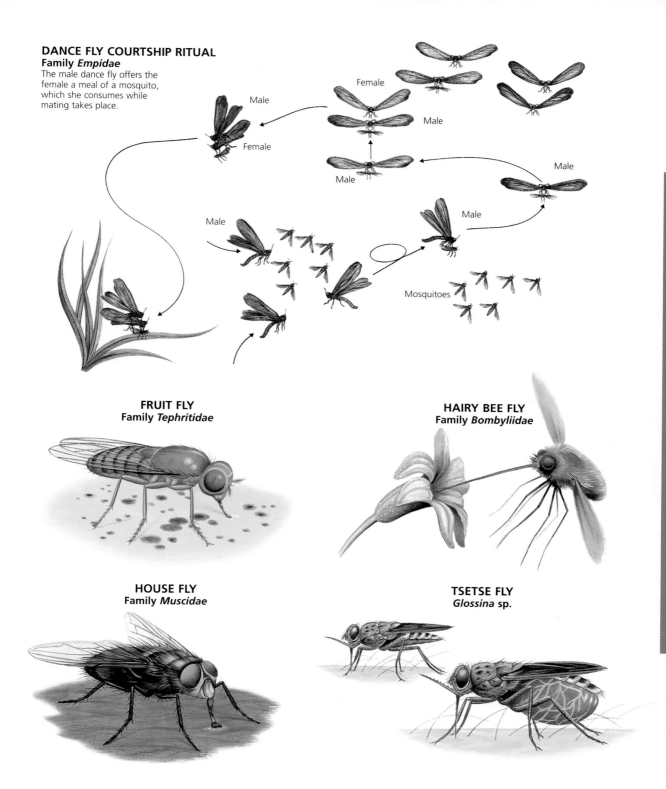

DANCE FLY COURTSHIP RITUAL
Family *Empidae*
The male dance fly offers the female a meal of a mosquito, which she consumes while mating takes place.

Male

Female

Female

Male

Male

Male

Male

Male

Male

Male

Mosquitoes

FRUIT FLY
Family *Tephritidae*

HAIRY BEE FLY
Family *Bombyliidae*

HOUSE FLY
Family *Muscidae*

TSETSE FLY
Glossina sp.

Butterflies and Moths

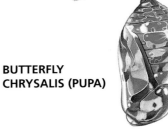

BUTTERFLY CHRYSALIS (PUPA)

BUTTERFLY AND MOTH COMPARISON

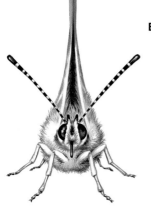

Butterfly wings fold
vertically when resting.

Moth wings are held
horizontal when resting.

Thin, threadlike antennae
with clubbed tips

Feathery or straight antennae
without clubbed tips

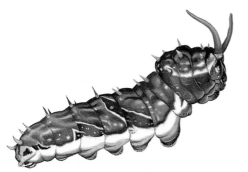

Butterfly caterpillar

Moth caterpillar

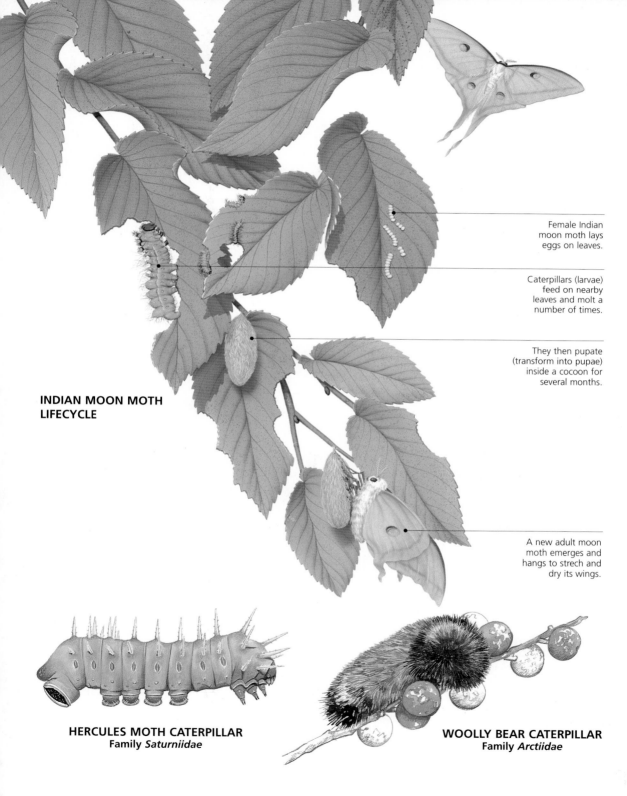

Female Indian moon moth lays eggs on leaves.

Caterpillars (larvae) feed on nearby leaves and molt a number of times.

They then pupate (transform into pupae) inside a cocoon for several months.

INDIAN MOON MOTH LIFECYCLE

A new adult moon moth emerges and hangs to strech and dry its wings.

HERCULES MOTH CATERPILLAR
Family *Saturniidae*

WOOLLY BEAR CATERPILLAR
Family *Arctiidae*

Butterflies

HACKBERRY BUTTERFLY
Family *Nymphalidae*

SWALLOWTAIL
Family *Papilionidae*

PAINTED LADY BUTTERFLY
Family *Nymphalidae*

PIPEVINE SWALLOWTAIL
Family *Papilionidae*

Anise swallowtail drinking nectar

Butterfly in flight

Butterfly at rest

DEADLY OR HARMLESS?

The harmless viceroy mimics the monarch so predators stay away.

The poisonous monarch butterfly has bold orange coloring as a warning.

88 BUTTERFLY
Diaethria sp.

MALAYAN LACEWING
BUTTERFLY
Family *Nymphalidae*

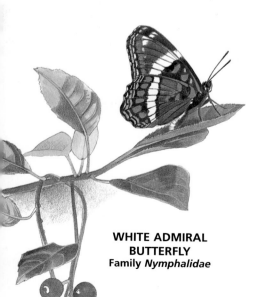

WHITE ADMIRAL
BUTTERFLY
Family *Nymphalidae*

BORDERED PATCH
BUTTERFLY
Family *Nymphalidae*

Moths

INDIAN MOON MOTH
Family *Saturniidae*

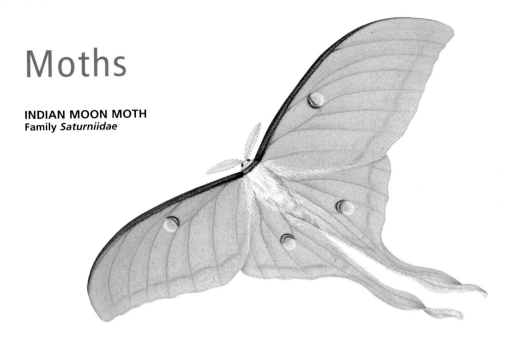

WHITE-LINED SPHINX MOTH
Family *Sphingidae*

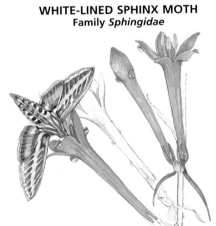

HUMMINGBIRD MOTH
Family *Sphingidae*

Moth larva eating leaves

Mexican 'jumping beans' (moth larvae)

Sheep moth

FIVE-SPOTTED BURNET MOTH
Family *Zygaenidae*

VAPORER MOTH CATERPILLAR
Family *Lymantriidae*

CERISY'S SPHINX MOTH
Family *Sphingidae*

YELLOW EMPEROR MOTH
Family *Saturniidae*

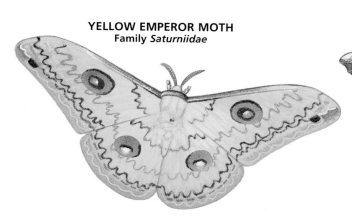

PUSS MOTH CATERPILLAR
Family *Notodontidae*

MADAGASCAN SUNSET MOTH
Family *Uraniidae*

Bees, Wasps and Ants

SAND WASP
Family *Sphecidae*

BULLDOG ANT
Family *Formicidae*

BUMBLEBEE
Family *Apidae*

POTTER WASP
Family *Vespidae*

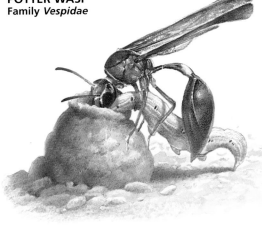

BUILDING A WASP NEST

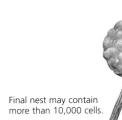

Solitary queen begins nest.　　She adds cells to lay eggs in.

First eggs hatch into workers,
who continue to build cells.

Final nest may contain
more than 10,000 cells.

Honey bee head

Velvet ant (species of wasp)

Army ant

Hairy-legged mining bee

Digger wasp gathering nectar

Trap-jaw ant head

ANTS AND PLANTS

Bull's horn acacia
has broad,
hollow thorns.

Some species of
ants use them
as their home.

In return for this
hospitality, the ants
chase off possible
threats, such as
beetles and cows.

QUEENLY DUTIES

Only a queen ant is able to
lay eggs, which are taken
by a worker and tended
until adulthood.

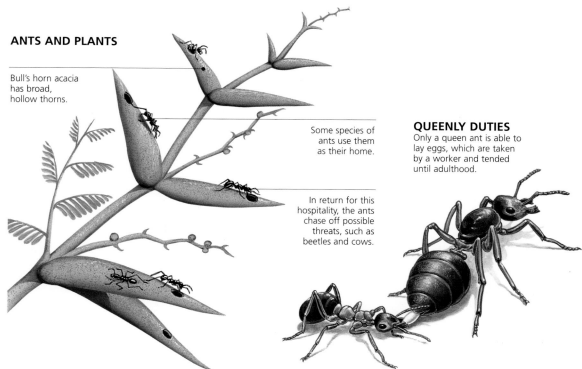

Wasps and Ants

EUROPEAN WASP
Family *Vespidae*

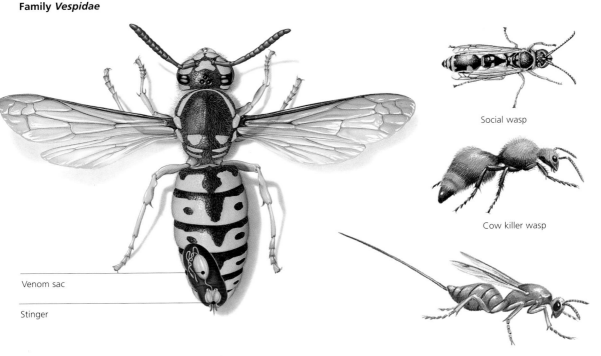

Venom sac

Stinger

ICHNEUMON WASP
Family *Ichneumonidae*

Female drills into a tree branch with her ovipositor (tubelike tail).

She stings a wood wasp larva, then deposits her eggs. These will feed on the larva when hatched.

TYPES OF WASPS

Social wasp

Cow killer wasp

Torymid wasp

Spider wasp

Mud dauber

TYPES OF BIG-HEADED ANTS

Worker

Soldier

Male

Queen

RED ANTS GREETING
Family *Formicidae*

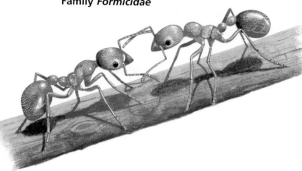

AMAZON ANT
Family *Formicidae*

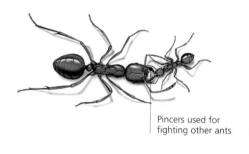

Pincers used for fighting other ants

LEAF-CUTTER ANTS
Atta sp.

Inside the nest, the leaves are chewed to a pulp, then used as compost for the special fungus the ants eat.

Small workers remove any parasitic flies from the leaves.

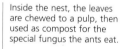

The leaf pieces are left at the entrance to be carried into the nest.

Medium-sized workers cut pieces of leaves to carry back to the nest.

Honey Bees

HONEY BEE DANCE

Worker bees returning from a food source dance to inform others in the hive where it is.

If the food source is close by, the dance is circular, the alignment showing the bee's direction relative to the sun.

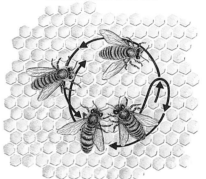

If the source is further away, the dance is a figure-of-eight, the speed and number of 'waggles' at the center indicating distance.

GATHERING NECTAR

Bees gather nectar from flowers to provide food for the hive, storing it on their legs.

Providing nectar for bees is an effective way for plants to have pollen distributed to fertilize other flowers.

After converting the nectar into honey, bees store it in cells for the hive to share.

INSIDE A BEE HIVE

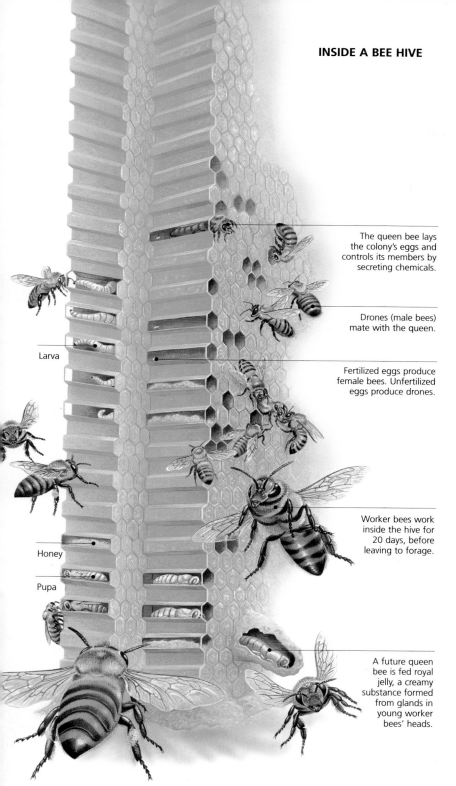

Larva

Honey

Pupa

The queen bee lays the colony's eggs and controls its members by secreting chemicals.

Drones (male bees) mate with the queen.

Fertilized eggs produce female bees. Unfertilized eggs produce drones.

Worker bees work inside the hive for 20 days, before leaving to forage.

A future queen bee is fed royal jelly, a creamy substance formed from glands in young worker bees' heads.

BEE EGG
Single egg laid in each cell

MOVING HIVES
Queen gathers a group of workers to leave the old hive.

The vulnerable queen is protected until a new home is found.

Scouts inform the others when a new home is found, like this beekeeper's box.

Classifying Animals

LEVELS OF CLASSIFICATION

Kingdom
There are five kingdoms, of which animals is one. Plants, fungi, protists and monerans are the others.

Phylum
First subdivision within the kingdom, linking broadly similar creatures. Vertebrates (chordata) form one phylum.

Class
Subdivision within the phylum linking similar orders. Mammals, birds, reptiles, amphibians and fishes are all classes.

Order
Subdivision within the class linking family groups by common characteristics. Marsupials and bats are orders.

Family
Subdivision within the order linking closely related individuals according to biology and behavior, like vipers or geckos.

Genus
Subdivision within the family, containing one or more species that are biologically very closely related, like salmons or tunas.

Species
Individuals within the genus that can interbreed and produce offspring that can interbreed, such as ostriches.

BROWN BEAR
Ursus arctos

ANIMALIA
(Kingdom)

CHORDATA
(Phylum)

MAMMALIA
(Class)

CARNIVORA
(Order)

URSIDAE
(Family)

URSUS
(Genus)

ARCTOS
(Species)

FIRE SALAMANDER
Salamandra salamandra

ANIMALIA
(Kingdom)

CHORDATA
(Phylum)

AMPHIBIA
(Class)

CAUDATA
(Order)

SALAMANDRIDAE
(Family)

SALAMANDRA
(Genus)

SALAMANDRA
(Species)

Mammals

CLASS MAMMALIA MAMMALS

SUBCLASS PROTOTHERIA EGG-LAYING MAMMALS
ORDER MONOTREMATA MONOTREMES

Tachyglossidae	Echidnas
Ornithorhynchidae	Duck-billed platypus

SUBCLASS THERIA MODERN MAMMALS
INFRACLASS METATHERIA POUCHED MAMMALS
ORDER MARSUPIALIA MARSUPIALS

Didelphidae	American opossums

Microbiotheriidae	Colocolos
Caenolestidae	Shrew-opossums
Dasyuridae	Marsupial mice
Myrmecobiidae	Numbat
Thylacinidae	Thylacine
Notoryctidae	Marsupial mole
Peramelidae	Bandicoots
Peroryctidae	Spiny bandicoots
Vombatidae	Wombats
Phascolarctidae	Koala
Phalangeridae	Phalangers
Petauridae	Gliding phalangers
Pseudocheiridae	Ringtail possums
Burramyidae	Pygmy possums
Acrobatidae	Feathertails
Tarsipedidae	Honey possum
Macropodidae	Kangaroos, wallabies
Potoroidae	Bettongs

INFRACLASS EUTHERIA PLACENTAL MAMMALS
ORDER XENARTHRA ANTEATERS, SLOTHS AND ARMADILLOS

Myrmecophagidae	American anteaters
Bradypodidae	Three-toed sloths
Megalonychidae	Two-toed sloths
Dasypodidae	Armadillos

ORDER PHOLIDOTA PANGOLINS, SCALY ANTEATERS

Manidae	Pangolins, scaly anteater

ORDER INSECTIVORA INSECTIVORES

Solenodontidae	Solenodons
Tenrecidae	Tenrecs, otter shrews
Chrysochloridae	Golden moles
Erinaceidae	Hedgehogs, moonrats
Soricidae	Shrews
Talpidae	Moles, desmans

ORDER SCANDENTIA TREE SHREWS

Tupaiidae	Tree shrews

ORDER DERMOPTERA FLYING LEMURS (COLUGOS)

Cynocephalidae	Flying lemurs, colugos

ORDER CHIROPTERA BATS

Pteropodidae	Old World fruit bats
Rhinopomatidae	Mouse-tailed bats
Emballonuridae	Sheath-tailed bats
Craseonycteridae	Hog-nosed bat, bumblebee bat
Nycteridae	Slit-faced bats
Megadermatidae	False vampire bats
Rhinolophidae	Horseshoe bats
Hipposideridae	Old World leaf-nosed bats
Noctilionidae	Bulldog bats

Mormoopidae	Naked-backed bats
Phyllostomidae	New World leaf-nosed bats
Natalidae	Funnel-eared bats
Furipteridae	Smoky bats
Thyropteridae	Disk-winged bats
Myzopodidae	Old World sucker-footed bats
Vespertilionidae	Vespertilionid bats
Mystacinidae	New Zealand short-tailed bats
Molossidae	Free-tailed bats

ORDER CETACEA — WHALES, DOLPHINS AND PORPOISES

Platanistidae	River dolphins
Delphinidae	Dolphins
Phocoenidae	Porpoises
Monodontidae	Narwhal, white whale
Physeteridae	Sperm whales
Ziphiidae	Beaked whales
Eschrichtiidae	Gray whale
Balaenopteridae	Rorquals
Balaenidae	Right whales

ORDER PRIMATES — PRIMATES

Cheirogaleidae	Dwarf lemurs
Lemuridae	Large lemurs
Megaladapidae	Sportive lemurs
Indridae	Leaping lemurs
Daubentoniidae	Aye-aye
Loridae	Lorises, galagos
Tarsiidae	Tarsiers
Callitrichidae	Marmosets, tamarins
Cebidae	New World monkeys
Cercopithecidae	Old World monkeys
Hylobatidae	Gibbons
Hominidae	Great apes, humans

ORDER SIRENIA — SEA COWS

Dugongidae	Dugong
Trichechidae	Manatees

ORDER PROBOSCIDEA — ELEPHANTS

Elephantidae	Elephants

ORDER PERISSODACTYLA — ODD-TOED UNGULATES

Equidae	Horses
Tapiridae	Tapirs
Rhinocerotidae	Rhinoceroses

ORDER CARNIVORA — CARNIVORES

Canidae	Dogs, foxes
Ursidae	Bears, pandas
Procyonidae	Racoons and relatives
Mustelidae	Mustelids
Viverridae	Civets and relatives
Herpestidae	Mongooses
Hyaenidae	Hyenas
Felidae	Cats
Otariidae	Sealions
Odobenidae	Walrus
Phocidae	Seals

ORDER HYRACOIDEA — HYRAXES

Procaviidae	Hyraxes

ORDER TUBULIDENTATA — AARDVARK

Orycteropodidae	Aardvark

ORDER ARTIODACTYLA   EVEN-TOED UNGULATES

Suidae	Pigs
Tayassuidae	Peccaries
Hippopotamidae	Hippopotamuses
Camelidae	Camels, camelids
Tragulidae	Mouse deer
Moschidae	Musk deer
Cervidae	Deer

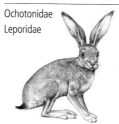

ORDER ARTIODACTYLA   EVEN-TOED UNGULATES

Giraffidae	Giraffe, okapi
Antilocapridae	Pronghorn
Bovidae	Cattle, antelopes, sheep and goats

ORDER RODENTIA   RODENTS

Aplodontidae	Mountain beaver
Sciuridae	Squirrels, marmots and relatives
Geomyidae	Pocket gophers
Heteromyidae	Pocket mice
Castoridae	Beavers
Anomaluridae	Scaly-tailed squirrels
Pedetidae	Spring hare
Muridae	Rats, mice, gerbils and relatives
Gliridae	Dormice
Seleviniidae	Desert dormouse
Zapodidae	Jumping mice
Dipodidae	Gerboas
Hystricidae	Old World porcupines
Erethizontidae	New World porcupines
Caviidae	Cavies and relatives
Hydrochaeridae	Capybara
Dinomyidae	Pacarana
Dasyproctidae	Agoutis, pacas
Chinchillidae	Chinchillas and relatives
Capromyidae	Hutias and relatives
Myocastoridae	Coypu

Octodontidae	Degus and relatives
Ctenomyidae	Tuco-tucos
Abrocomidae	Chinchilla-rats
Echymidae	Spiny rats
Thryonomyidae	Cane rats
Petromyidae	African rock-rat
Bathyergidae	African mole-rats
Ctenodactylidae	Gundis

ORDER LAGOMORPHA   LAGOMORPHS

Ochotonidae	Pikas
Leporidae	Rabbits, hares

ORDER MACROSCELIDEA   ELEPHANT SHREWS

Macroscelididae	Elephant shrews

Birds

CLASS AVES — BIRDS

ORDER STRUTHIONIFORMES — RATITES AND TINAMOUS

Struthionidae	Ostrich
Tinamidae	Tinamous
Rheidae	Rheas
Casuariidae	Cassowaries
Dromaiidae	Emu
Apterygidae	Kiwis

ORDER PROCELLARIIFORMES — ALBATROSSES AND PETRELS

Diomedeidae	Albatrosses
Procellariidae	Shearwaters
Hydrobatidae	Storm petrels
Pelecanoididae	Diving petrels

ORDER SPHENISCIFORMES — PENGUINS

Spheniscidae	Penguins

ORDER GAVIIFORMES — DIVERS

Gaviidae	Divers (loons)

ORDER PODICIPEDIFORMES — GREBES

Podicipedidae	Grebes

ORDER PELECANIFORMES — PELICANS AND RELATIVES

Phaethontidae	Tropicbirds
Pelecanidae	Pelicans
Phalacrocoracidae	Cormorants, anhingas
Sulidae	Gannets, boobies
Fregatidae	Frigatebirds

ORDER CICONIIFORMES — HERONS AND RELATIVES

Ardeidae	Herons, bitterns
Scopidae	Hammerhead
Ciconiidae	Storks
Balaenicipitidae	Whale-headed stork
Threskiornithidae	Ibises, spoonbills

ORDER PHOENICOPTERIFORMES — FLAMINGOS

Phoenicopteridae	Flamingos

ORDER FALCONIFORMES — RAPTORS

Accipitridae	Osprey, kites, hawks, eagles, Old World vultures, harriers, buzzards, harpies, buteonines
Sagittariidae	Secretarybird
Falconidae	Falcons, falconets, caracaras
Cathartidae	New World vultures

ORDER ANSERIFORMES — WATERFOWL AND SCREAMERS

Anatidae	Geese, swans, ducks
Anhimidae	Screamers

ORDER GALLIFORMES — GAMEBIRDS

Megapodiidae	Megapodes (mound-builders)
Cracidae	Chachalacas, guans, curassows
Phasianidae	Turkeys, grouse and relatives

ORDER OPISTHOCOMIFORMES — HOATZIN

Opisthocomidae	Hoatzin

ORDER GRUIFORMES — CRANES AND RELATIVES

Mesitornithidae	Mesites
Turnicidae	Hemipode-quails (button quails)
Pedionomidae	Collared hemipode
Gruidae	Cranes
Aramidae	Limpkins
Psophiidae	Trumpeters
Rallidae	Rails
Heliornithidae	Finfoots
Rhynochetidae	Kagus
Eurypygidae	Sunbittern
Cariamidae	Seriemas
Otididae	Bustards

ORDER CHARADRIIFORMES — WADERS AND SHOREBIRDS

Jacanidae	Jacanas
Rostratulidae	Painted snipe
Dromadidae	Crab plover
Haematopodidae	Oystercatchers
Ibidorhynchidae	Ibisbill
Recurvirostridae	Stilts, avocets
Burhinidae	Stone curlews (thick knees)
Glareolidae	Coursers, pratincoles

Charadriidae	Plovers, dotterels
Scolopacidae	Curlews, sandpipers, snipes
Thinocoridae	Seedsnipes
Chionididae	Sheathbills
Laridae	Gulls, terns, skimmers
Stercorariidae	Skuas, jaegers
Alcidae	Auks

ORDER COLUMBIFORMES — PIGEONS AND SANDGROUSE

Pteroclididae	Sandgrouse
Columbidae	Pigeons, doves

ORDER PSITTACIFORMES — PARROTS

Cacatuidae	Cockatoos
Psittacidae	Parrots

ORDER CUCULIFORMES — TURACOS AND CUCKOOS

Musophagidae	Turacos, louries (plantain-eaters)
Cuculidae	Cuckoos and relatives

ORDER STRIGIFORMES — OWLS

Tytonidae	Barn owls, bay owls
Strigidae	Hawk owls (true owls)

ORDER CAPRIMULGIFORMES — NIGHTJARS AND FROGMOUTHS

Steatornithidae	Oilbird
Podargidae	Frogmouths
Nyctibiidae	Potoos
Aegothelidae	Owlet nightjars
Caprimulgidae	Nightjars

ORDER APODIFORMES — SWIFTS AND HUMMINGBIRDS

Apodidae	Swifts
Hemiprocnidae	Crested swifts
Trochilidae	Hummingbirds

ORDER COLIIFORMES — MOUSEBIRDS

Coliidae	Mousebirds

ORDER TROGONIFORMES — TROGONS

Trogonidae	Trogons

ORDER CORACIIFORMES KINGFISHERS AND RELATIVES

Alcedinidae	Kingfishers
Todidae	Todies
Momotidae	Motmots
Meropidae	Bee-eaters
Coraciidae	Rollers
Upupidae	Hoopoe
Phoeniculidae	Wood-hoopoes
Bucerotidae	Hornbills

ORDER PICIFORMES WOODPECKERS AND BARBETS

Galbulidae	Jacamars
Bucconidae	Puffbirds
Capitonidae	Barbets
Ramphastidae	Toucans
Indicatoridae	Honeyguides
Picidae	Woodpeckers

ORDER PASSERIFORMES PASSERINES OR PERCHING BIRDS

Suborder Eurylaimi   Broadbills and Pittas

Eurylaimidae	Broadbills
Philepittidae	Sunbirds, asitys
Pittidae	Pittas
Acanthisittidae	New Zealand wrens

Suborder Furnarii   Ovenbirds and relatives

Dendrocolaptidae	Woodcreepers
Furnariidae	Ovenbirds
Formicariidae	Antbirds
Rhinocryptidae	Tapaculos

Suborder Tyranni   Tyrant Flycatchers and Relatives

Tyrannidae	Tyrant flycatchers
Pipridae	Manakins
Cotingidae	Cotingas
Oxyruncidae	Sharpbills
Phytotomidae	Plantcutters

Suborder Oscines   Songbirds

Menuridae	Lyrebirds

Atrichornithidae	Scrub-birds
Alaudidae	Larks
Motacillidae	Wagtails, pipits
Hirundinidae	Swallows, martins
Campephagidae	Cuckoo-shrikes and relatives
Pycnonotidae	Bulbuls
Irenidae	Leafbirds, ioras, bluebirds
Laniidae	Shrikes
Vangidae	Vangas
Bombycillidae	Waxwings
Hypocoliidae	Hypocolius
Ptilogonatidae	Silky flycatchers
Dulidae	Palmchat
Prunellidae	Accentors, hedge-sparrows
Mimidae	Mockingbirds and relatives
Cinclidae	Dippers
Turdidae	Thrushes
Timaliidae	Babblers and relatives
Troglodytidae	Wrens
Sylviidae	Old World warblers
Muscicapidae	Old World flycatchers
Maluridae	Fairywrens and relatives
Acanthizidae	Australian warblers and relatives
Ephthianuridae	Australian chats
Orthonychidae	Logrunners and relatives
Rhipiduridae	Fantails
Monarchidae	Monarch flycatchers
Petroicidae	Australasian robins
Pachycephalidae	Whistlers and relatives
Aegithalidae	Long-tailed tits
Remizidae	Penduline tits

Paridae	True tits, chickadees, titmice
Sittidae	Nuthatches, sitellas, wallcreeper
Certhiidae	Holarctic treecreepers
Rhabdornithidae	Philippine treecreepers
Climacteridae	Australasian treecreepers
Dicaeidae	Flowerpeckers, pardalotes
Nectariniidae	Sunbirds
Zosteropidae	White-eyes
Meliphagidae	Honeyeaters
Vireonidae	Vireos
Emberizidae	Buntings, tanagers
Parulidae	New World wood warblers
Icteridae	Icterids (American blackbirds)
Fringillidae	Finches

Drepanididae	Hawaiian honeycreepers
Estrildidae	Estrildid finches
Ploceidae	Weavers
Passeridae	Old World sparrows
Sturnidae	Starlings, mynahs
Oriolidae	Orioles, figbirds
Dicruridae	Drongos
Callaeidae	New Zealand wattlebirds
Grallinidae	Magpie-larks
Artamidae	Wood swallows
Cracticidae	Bell magpies
Ptilonorhynchidae	Bowerbirds
Paradisaeidae	Birds-of-paradise
Corvidae	Crows, jays

Reptiles

CLASS REPTILIA   REPTILES

SUBCLASS EUREPTILIA

SUPERORDER TESTUDINE
ORDER TESTUDINATA   TURTLES, TERRAPINS AND TORTOISES

Suborder Pleurodira	**Side-neck Turtles**
Chelidae	Snake-neck turtles
Pelomedusidae	Helmeted side-neck turtles

Suborder Cryptodira	**Hidden-necked Turtles**
Superfamily Trionychoidea	
Kinosternidae	Mud and musk turtles
Dermatemydidae	Mesoamerican river turtle
Carretochelyidae	Australian softshell turtle
Trionychidae	Holarctic and paleotropical softshell turtles

Superfamily Cheloniodea	
Dermochelyidae	Leatherback sea turtles
Cheloniidae	Sea turtles

Superfamily Testudinoidea	
Chelydridae	Snapping turtles
Emydidae	New World pond turtles and terrapins
Testudinidae	Tortoises
Bataguridae	Old World pond turtles

SUPERORDER ARCHOSAURIA
ORDER CROCODILIA   CROCODILIANS

Alligatoridae	Alligators and caimans
Crocodylidae	Crocodiles
Gavialidae	Gharials

SUPERORDER LEPIDOSAURIA

ORDER RHYNCHOCEPHALIA	TUATARA
Sphenodontidae	Tuatara

ORDER SQUAMATA	SQUAMATES

Suborder Amphisbaenia — Amphisbaenians

Bipedidae	Ajolotes
Amphisbaenidae	Worm lizards
Trogonophidae	Desert ringed lizards
Rhineuridae	Florida worm lizard

Suborder Iguania — Iguanid Lizards

Corytophanidae	Helmeted lizards
Crotaphytidae	Collared and leopard lizards
Hoplocercidae	Hoplocercids
Iguanidae	Iguanas

Opluridae	Madagascar iguanians
Phrynosomatidae	Scaly, sand and horned lizards
Polychrotidae	Anoloid lizards
Tropiduridae	Tropidurids
Agamidae	Agamid lizards
Chameleonidae	Chameleons

Suborder Scleroglossa

Superfamily Gekkonoidea

Eublepharidae	Eye-lash geckos
Gekkonidae	Geckos
Pygopodidae	Australasian flapfoots

Superfamily Scincoidea

Xantusiidae	Night lizards
Lacertidae	Lacertids
Scincidae	Skinks
Dibamidae	Dibamids
Cordylidae	Girdle-tailed lizards
Gerrhosauridae	Plated lizards
Teiidae	Macroteiids
Gymnophthalmidae	Microteiids

Superfamily Anguoidea

Xenosauridae	Knob-scaled lizards
Anguidae	Anguids; glass and alligator lizards
Helodermatidae	Beaded lizards
Varanidae	Monitor lizards
Lanthanotidae	Earless monitor lizard

Suborder Serpentes — Snakes

Infraorder Scolecophidia

Anomalepididae	Blind wormsnakes
Typhlopidae	Blind snakes
Leptotyphlopidae	Thread snakes

Infraorder Alethinophidia

Anomochelidae	Stump heads
Aniliidae	Coral pipesnakes
Cylindrophidae	Asian pipesnakes
Uropeltidae	Shield tails
Xenopeltidae	Sunbeam snake
Loxocemidae	Dwarf boa
Boidae	Pythons and boas

Ungaliophiidae	Ungaliophiids
Bolyeriidae	Round Island snakes
Tropidophiidae	Woodsnakes
Acrochordidae	File snakes
Atractaspididae	Mole vipers
Colubridae	Harmless and rear-ranged snakes
Elapidae	Cobras, kraits, coral snakes and sea snakes
Viperidae	Adders and vipers

Amphibians

CLASS AMPHIBIA   AMPHIBIANS

SUBCLASS LISSAMPHIBIA
ORDER CAUDATA   SALAMANDERS AND NEWTS

Suborder Sirenoidea
Sirenidae — Sirens

Suborder Cryptobranchoidea
Cryptobranchidae — Hellbenders and giant salamanders
Hynobiidae — Hynobiids

Suborder Salamandroidea
Amphiumidae — Amphiumas (congo eels)
Plethodontidae — Lungless salamanders
Rhyacotritonidae — Torrent salamanders
Proteidae — Mudpuppies, waterdogs, and the olm
Salamandridae — Salamandrids

Ambystomatidae — Mole salamanders
Dicamptodontidae — Dicamptodontids

ORDER GYMNOPHIONA   CAECILIANS
Rhinatrematidae — South American tailed caecilians
Ichthyophiidae — Ichthyophiids
Uraeotyphlidae — Uraeotyphlids
Scolecomorphidae — Scolecomorphids
Caeciliidae — Caeciliids and aquatic caecilians

ORDER ANURA   FROGS AND TOADS
Ascaphidae — "Tailed" frogs
Leiopelmatidae — New Zealand frogs
Bombinatoridae — Fire-bellied toads and allies

Discoglossidae — Discoglossid frogs
Pipidae — Pipas and "clawed" frogs
Rhinophrynidae — Cone-nosed frog
Megophryidae — Megophryids
Pelodytidae — Parsley frogs
Pelobatidae — Spadefoots

Superfamily Bufonoidea
Allophrynidae — Allophrynid frog
Brachycephalidae — Saddleback frogs
Bufonidae — Toads
Helophrynidae — Ghost frogs
Leptodactylidae — Neotropical frogs
Myobatrachidae — Australasian frogs
Sooglossiidae — Seychelles frogs
Rhinodermatidae — Darwin's frogs
Hylidae — Hylid treefrogs
Pelodryadidae — Australasian treefrogs
Centrolenidae — Glass frogs
Pseudidae — Natator frogs
Dendrobatidae — Dart-poison frogs

Superfamily Ranoidea
Microhylidae — Microhylids
Hemisotidae — Shovel-nosed frogs
Arthroleptidae — Squeakers
Ranidae — Ranid frogs
Hyperoliidae — Reed and lily frogs
Rhacophoridae — Rhacaphorid treefrogs

Fishes

SUPERCLASS AGNATHA — JAWLESS FISHES

CLASS MYXINI — HAGFISHES

ORDER MYXINIFORMES	HAGFISHES
Myxinidae	Hagfishes

CLASS CEPHALASPIDOMORPHI LAMPREYS AND RELATIVES

ORDER PETROMYZONTIFORMES	LAMPREYS
Petromyzontidae	Northern lampreys
Geotriidae	Pouched lampreys
Mordaciidae	Shorthead lampreys

SUPERCLASS GNATHOSTOMATA — JAWED FISHES

CLASS CHONDRICHTHYES CARTILAGINOUS FISHES

SUBCLASS ELASMOBRANCHII	SHARKS AND RAYS
ORDER HEXANCHIFORMES	SIX- AND SEVENGILL SHARKS
Chlamydoselachidae	Frill shark
Hexanchidae	Sixgill, sevengill (cow) sharks

ORDER SQUALIFORMES	DOGFISH SHARKS AND RELATIVES
Echinorhinidae	Bramble sharks
Squalidae	Dogfish sharks
Oxynotidae	Roughsharks

ORDER PRISTIOPHORIFORMES	SAWSHARKS
Pristiophoridae	Sawsharks

ORDER HETERODONTIFORMES	HORN (BULLHEAD) SHARKS
Heterodontidae	Horn (bullhead) sharks

ORDER ORECTOLOBIFORMES	CARPETSHARKS AND RELATIVES
Parascylliidae	Collared carpetsharks
Brachaeluridae	Blind sharks
Orectolobidae	Wobbegongs
Hemiscylliidae	Longtail carpetsharks
Stegostomatidae	Zebra shark
Ginglymostomatidae	Nurse sharks
Rhincodontidae	Whale shark

ORDER LAMNIFORMES	MACKEREL SHARKS AND RELATIVES
Odontaspididae	Sand tigers (gray nurse sharks)
Mitsukurinidae	Goblin shark
Pseudocarchariidae	Crocodile shark
Megachasmidae	Megamouth shark
Alopiidae	Thresher sharks
Cetorhinidae	Basking shark
Lamnidae	Mackerel sharks (makos)

ORDER CARCHARHINIFORMES	GROUND SHARKS
Scyliorhinidae	Catsharks
Proscylliidae	Finback catsharks
Pseudotriakidae	False catshark
Leptochariidae	Barbled houndshark
Triakidae	Houndsharks
Hemigaleidae	Weasel sharks
Carcharhinidae	Requiem (whaler) sharks
Sphyrnidae	Hammerhead sharks

Classifying Animals

FACT FILE

ORDER SQUATINIFORMES ANGEL SHARKS

Squatinidae	Angel sharks

ORDER RHINOBATIFORMES — SHOVELNOSE RAYS

Platyrhinidae	Platyrhinids
Rhinobatidae	Guitarfishes
Rhynchobatidae	Sharkfin guitarfishes

ORDER RAJIFORMES — SKATES

Rajidae	Skates

ORDER PRISTIFORMES — SAWFISHES

Pristidae	Sawfishes

ORDER TORPEDINIFORMES — ELECTRIC RAYS

Torpedinidae	Electric rays
Hypnidae	Coffinrays
Narcinidae	Narcinids
Narkidae	Narkids

ORDER MYLIOBATIFORMES — STINGRAYS AND RELATIVES

Potamotrygonidae	River rays
Dasyatididae	Stingrays
Urolophidae	Round stingrays
Gymnuridae	Butterfly rays
Hexatrygonidae	Sixgill rays
Myliobatididae	Eagle rays
Rhinopteridae	Cownose rays
Mobulidae	Mantas, devil rays

SUBCLASS HOLOCEPHALI — CHIMAERAS
ORDER CHIMAERIFORMES — CHIMAERAS

Callorhynchidae	Plownose chimaeras (Elephant fishes)
Chimaeridae	Shortnose chimaeras (Ghost sharks, ratfishes)
Rhinochimaeridae	Longnose chimaeras (spookfishes)

CLASS OSTEICHTHYES — BONY FISHES

SUBCLASS SARCOPTERYGII — LUNGFISHES AND COELACANTH (FLESHY-FINNED FISHES)

ORDER CERATODONTIFORMES — LUNGFISHES

Ceratodontidae	Australian lungfish

ORDER LEPIDOSIRENIFORMES — LUNGFISHES

Lepidosirenidae	South American lungfish
Protopteridae	African lungfishes

ORDER COELACANTHIFORMES — COELACANTH

Latimeriidae	Coelacanth (gombessa)

SUBCLASS ACTINOPTERYGII — RAY-FINNED FISHES
ORDER POLYPTERIFORMES BICHIRS

Polypteridae	Bichirs

ORDER ACIPENSERIFORMES — STURGEONS AND RELATIVES

Acipenseridae	Sturgeons
Polyodontidae	Paddlefishes

ORDER LEPISOSTEIFORMES GARS

Lepisosteidae	Gars

ORDER AMIIFORMES BOWFIN

Amiidae	Bowfin

GROUP TELEOSTEI   TELEOSTS

ORDER HIODONTIFORMES   MOONEYES

Hiodontidae   Mooneyes

ORDER OSTEOGLOSSIFORMES   BONYTONGUES AND RELATIVES

Osteoglossidae   Bonytongues
Pantodontidae   Freshwater butterflyfish
Notopteridae   Featherbacks
Mormyridae   Elephantfishes
Gymnarchidae   Aba

ORDER ELOPIFORMES   TARPONS AND RELATIVES

Elopidae   Tenpounders (ladyfishes)
Megalopidae   Tarpons

ORDER ALBULIFORMES   BONEFISHES AND RELATIVES

Albulidae   Bonefishes

ORDER NOTACANTHIFORMES   SPINY EELS AND RELATIVES

Halosauridae   Halosaurs
Notacanthidae   Spiny eels

ORDER ANGUILLIFORMES   EELS

Anguillidae   Freshwater eels
Heterenchelyidae   Shortfaced eels
Moringuidae   Spaghetti eels
Chlopsidae   False moray eels
Myrocongridae   Myrocongrid
Muraenidae   Moray eels
Nemichthyidae   Snipe eels
Muraenesocidae   Pike eels
Synaphobranchidae   Cutthroat eels
Ophichthidae   Snake eels, worm eels
Nettastomatidae   Duckbill eels
Colocongridae   Colocongrids
Congridae   Conger eels, garden eels
Derichthyidae   Narrowneck eels, spoonbill eels
Serrivomeridae   Sawtooth eels
Cyematidae   Bobtail snipe eels

Saccopharyngidae   Swallowers
Eurypharyngidae   Gulper eels
Monognathidae   Singlejaw eels

ORDER CLUPEIFORMES   SARDINES, HERRINGS, AND ANCHOVIES

Denticipitidae   Denticle herring
Clupeidae   Sardines, herrings, pilchards
Engraulididae   Anchovies
Chirocentridae   Wolf herrings

ORDER GONORYNCHIFORMES   MILKFISH, BEAKED SALMONS AND RELATIVES

Chanidae   Milkfish
Gonorynchidae   Beaked salmons
Kneriidae   Shellears
Phractolaemidae   Hingemouth

ORDER CYPRINIFORMES   CARPS, MINNOWS AND RELATIVES

Cyprinidae   Carps, minnows
Psilorhynchidae   Psilorhynchids
Balitoridae   Hillstream loaches
Cobitidae   Loaches
Gyrinocheilidae   Algae eaters
Catostomidae   Suckers

ORDER CHARACIFORMES   CHARACINS AND RELATIVES

Citharinidae   Citharinids
Distichodontidae   Distichodontins, African fin eaters
Hepsetidae   African pike characin
Erythrinidae   Trahiras and relatives
Ctenoluciidae   American pike characins
Lebiasinidae   Voladoras, pencil fishes
Characidae   African tetras (characins), characins, tetras, brycons, piranhas and relatives
Curimatidae   Toothless characins
Prochilodontidae   Flannelmouth characins
Anostomidae   Anostomins, leporinins
Hemiodontidae   Hemiodontids
Chilodontidae   Headstanders
Gasteropelecidae   Freshwater hatchetfishes

ORDER SILURIFORMES — CATFISHES AND KNIFEFISHES

Suborder Siluroidei

Diplomystidae	Velvet catfishes
Nematogenyidae	Mountain catfish
Trichomycteridae	Spinyhead catfishes, candirus
Callichthyidae	Plated catfishes
Scoloplacidae	Spinynose catfishes
Astroblepidae	Andes catfishes
Loricariidae	Armored suckermouth catfishes
Ictaluridae	Bullhead catfishes, madtoms
Bagridae	Bagrid catfishes
Cranoglanididae	Armorhead catfishes
Siluridae	Sheatfishes, wels, glass catfishes
Schilbidae	Schilbid catfishes
Pangasiidae	Pangasiid catfishes
Amblycipitidae	Torrent catfishes
Amphiliidae	Loach catfishes
Akysidae	Asian banjo catfishes
Sisoridae	Hillstream catfishes
Clariidae	Labyrinth (airbeathing) catfishes
Heteropneustidae	Airsac catfishes
Chacidae	Square head (angler) catfishes
Olyridae	Olyrid catfishes
Malapteruridae	Electric catfishes
Ariidae	Sea catfishes
Plotosidae	Eeltail catfishes
Mochokidae	Upsidedown catfishes (squeakers)
Doradidae	Thorny catfishes
Auchenipteridae	Wood catfishes
Pimelodidae	Longwhisker, shovelnose catfishes
Ageneiosidae	Bottlenose (barbelless) catfishes
Helogenidae	Helogene catfishes
Cetopsidae	Carneros, shark (whale) catfishes
Hypophthalmidae	Loweye catfishes
Aspredinidae	Banjo catfishes

Suborder Gymnotoidei

Sternopygidae	Longtail knifefishes
Rhamphichthyidae	Longsnout knifefishes
Hypopomidae	Bluntnose knifefishes
Apternotidae	Black knifefishes
Gymnotidae	Banded knifefishes
Electrophoridae	Electric eel (electric knifefish)

ORDER ARGENTINIFORMES — HERRINGS, SMELTS, BARRELEYES AND RELATIVES

Argentinidae	Herring smelts
Microstomatidae	Deepsea smelts
Opisthoproctidae	Barreleyes (spookfishes)
Alepocephalidae	Slickheads
Bathylaconidae	Bathylaconids
Platytroctidae	Tubeshoulders

ORDER SALMONIFORMES — SALMONS, SMELTS AND RELATIVES

Osmeridae	Northern smelts, ayu
Salangidae	Icefishes (noodlefishes)
Retropinnidae	Southern smelts
Galaxiidae	Galaxiids
Lepidogalaxiidae	Salamanderfish
Salmonidae	Trouts, salmons, chars, whitefishes

ORDER ESOCIFORMES — PIKES, PICKERELS AND RELATIVES

Esocidae	Pikes
Umbridae	Mudminnows

ORDER ATELEOPODIFORMES — JELLYNOSE FISHES

Ateleopodidae	Jellynose fishes

ORDER STOMIIFORMES — DRAGONFISHES, LIGHTFISHES AND RELATIVES

Gonostomatidae	Lightfishes, bristlemouths
Sternoptychidae	Hatchetfishes and relatives
Phosichthyidae	Lightfishes
Stomiidae	Viperfishes, dragonfishes, snaggletooths, loosejaws

ORDER AULOPIFORMES	LIZARDFISHES AND RELATIVES
Aulopidae	Aulopids
Bathysauridae	Bathysaurids
Chlorophthalmidae	Greeneyes
Ipnopidae	Ipnopids, tripod fishes
Synodontidae	Lizardfishes, Bombay duck
Scopelarchidae	Pearleyes
Notosudidae	Waryfishes
Giganturidae	Telescope fishes
Paralepididae	Barracudinas
Anotopteridae	Daggertooth
Evermannellidae	Sabertooth fishes
Omosudidae	Omosudid
Alepisauridae	Lancetfishes
Pseudotrichonotidae	Pseudotrichonotids

ORDER MYCTOPHIFORMES	LANTERNFISHES AND RELATIVES
Neoscopelidae	Neoscopelids
Myctophidae	Lanternfishes

ORDER LAMPRIDIFORMES	OARFISHES AND RELATIVES
Lampridae	Opahs
Veliferidae	Velifers
Lophotidae	Crestfishes
Radiicephalidae	Inkfishes
Trachipteridae	Ribbonfishes
Regalecidae	Oarfishes
Stylephoridae	Tube eye

ORDER POLYMIXIIFORMES	BEARDFISHES
Polymixiidae	Beardfishes

ORDER PERCOPSIFORMES	TROUTPERCHES AND RELATIVES
Percopsidae	Troutperches
Aphredoderidae	Pirate perch
Amblyopsidae	Cavefishes

ORDER GADIFORMES	CODS, HAKES AND RELATIVES
Muraenolepididae	Muraenolepidids
Moridae	Morid cods (moras)
Melanonidae	Melanonids
Euclichthyidae	Euclichthyid
Bregmacerotidae	Codlets
Gadidae	Codfishes, haddocks and relatives
Merlucciidae	Hakes
Steindachneriidae	Steindachneriid
Macrouridae	Rattails (grenadiers)

ORDER OPHIDIIFORMES	CUSKEELS, PEARLFISHES AND RELATIVES
Ophidiidae	Cuskeels, brotulas and relatives
Carapidae	Pearlfishes
Bythitidae	Livebearing brotulas
Aphyonidae	Aphyonids

ORDER BATRACHOIDIFORMES	TOADFISHES AND MIDSHIPMEN
Batrachoididae	Toadfishes and midshipmen

ORDER LOPHIIFORMES	ANGLERFISHES, GOOSEFISHES AND FROGFISHES
Lophiidae	Goosefishes (monkfishes)
Antennariidae	Frogfishes (shallow anglerfishes)
Tetrabrachiidae	Humpback angler
Lophichthyidae	Lophichthyid
Brachionichthyidae	Warty anglers (hand fishes)
Chaunacidae	Seatoads (gapers, coffinfishes)
Ogcocephalidae	Batfishes
Caulophrynidae	Fanfin anglers
Ceratiidae	Seadevils
Gigantactinidae	Whipnose anglers
Neoceratiidae	Needlebeard angler
Linophrynidae	Netdevils
Oneirodidae	Dreamers
Thaumatichthyidae	Wolftrap angler
Centrophrynidae	Hollowchin anglers
Diceratiidae	Double anglers
Himantolophidae	Footballfishes
Melanocetidae	Blackdevils

ORDER GOBIESOCIFORMES
CLINGFISHES AND RELATIVES

Gobiesocidae	Clingfishes
Callionymidae	Dragonets
Draconettidae	Draconettids

ORDER CYPRINODONTIFORMES
KILLIFISHES AND RELATIVES

Rivulidae	South American annuals
Aplocheilidae	African annuals
Profundulidae	Profundulids
Fundulidae	Killifishes
Valenciidae	Valenciids
Anablepidae	Foureyed fishes (cuatro ojos)
Poeciliidae	Livebearers (guppies, etc.)
Goodeidae	Goodeids
Cyprinodontidae	Pupfishes

ORDER BELONIFORMES
RICEFISHES, FLYINGFISHES AND RELATIVES

Adrianichthyidae	Ricefishes
Exocoetidae	Flyingfishes
Hemiramphidae	Halfbeaks
Belonidae	Needlefishes
Scomberesocidae	Sauries

ORDER ATHERINIFORMES
SILVERSIDES, RAINBOWFISHES AND RELATIVES

Atherinidae	Silversides, topsmelts, grunions
Dentatherinidae	Dentatherinid
Notocheiridae	Surf silversides
Melanotaeniidae	Rainbowfishes
Pseudomugilidae	Blue eyes
Telmatherinidae	Sailfin silversides
Phallostethidae	Priapium fishes

ORDER STEPHANOBERYCIFORMES
PRICKLEFISHES AND RELATIVES

Stephanoberycidae	Pricklefishes
Melamphaidae	Bigscale fishes (ridgeheads)
Gibberichthyidae	Gibberfishes
Hispidoberycidae	Bristlyskin
Rondeletiidae	Orangemouth whalefishes
Barbourisiidae	Redvelvet whalefish
Cetomimidae	Flabby whalefishes
Mirapinnidae	Hairyfish, tapetails
Megalomycteridae	Mosaicscale (bignose) fishes

ORDER BERYCIFORMES
SQUIRRELFISHES AND RELATIVES

Holocentridae	Squirrelfishes (soldier fishes)
Berycidae	Alfoncinos
Anoplogasteridae	Fangtooth fishes
Monocentrididae	Pineapple (pinecone) fishes
Anomalopidae	Flashlight (lanterneye) fishes
Trachichthyidae	Roughies (slimeheads)
Diretmidae	Spinyfins

ORDER ZEIFORMES
DORIES AND RELATIVES

Parazenidae	Parazen
Macrurocyttidae	Macrurocyttids
Zeidae	Dories
Oreosomatidae	Oreos
Grammicolepididae	Tinselfishes
Caproidae	Boarfishes

ORDER GASTEROSTEIFORMES
STICKLEBACKS, PIPEFISHES AND RELATIVES

Hypoptychidae	Sand eel
Aulorhynchidae	Tubesnouts
Gasterosteidae	Sticklebacks
Indostomidae	Paradox fish
Pegasidae	Seamoths
Aulostomidae	Trumpetfishes
Fistulariidae	Cornetfishes
Macroramphosidae	Snipefishes
Centriscidae	Shrimpfishes (razorfishes)
Solenostomidae	Ghost pipefishes
Syngnathidae	Pipefishes, seahorses

ORDER SYNBRANCHIFORMES — SWAMPEELS AND RELATIVES

Synbranchidae	Swampeels
Mastacembelidae	Spiny eels
Chaudhuriidae	Chaudhuriids

ORDER SCORPAENIFORMES — SCORPIONFISHES AND RELATIVES

Scorpaenidae	Scorpionfishes (rockfishes), stonefishes

Caracanthidae	Coral crouchers
Aploactinidae	Velvetfishes
Pataecidae	Prowfishes
Gnathanacanthidae	Red velvetfish
Congiopodidae	Pigfishes (horsefishes)
Triglidae	Sea robins (gurnards)
Dactylopteridae	Helmet (flying) gurnards

Platycephalidae	Flatheads
Bembridae	Deepwater flatheads
Hoplichthyidae	Spiny (ghost) flatheads
Anoplopomatidae	Sablefish, skilfish
Hexagrammidae	Greenlings
Zaniolepididae	Combfishes
Normanichthyidae	Southern sculpin
Rhamphocottidae	Grunt sculpin
Ereuniidae	Deepwater sculpins
Cottidae	Sculpins
Comephoridae	Baikal oilfishes
Abyssocottidae	Abyssocottids
Hemitripteridae	Hemitripterids
Psychrolutidae	Flatheads (blob fishes)
Bathylutichthyidae	Bathylutichthyids
Agonidae	Poachers
Cyclopteridae	Lumpfishes
Liparidae	Snailfishes, lumpsuckers

ORDER PERCIFORMES — PERCHES AND RELATIVES

Suborder Percoidei

Ambassidae	Glassfishes
Acropomatidae	Acropomatids
Epigonidae	Epigonids
Scrombropidae	Gnomefishes
Symphysanodontidae	Symphsanodontids
Caesioscorpididae	Caesioscorpidids
Polyprionidae	Wreckfishes
Dinopercidae	Cavebasses
Centropomidae	Snooks
Latidae	Giant perches
Percichthyidae	South temperate basses
Moronidae	North temperate basses
Serranidae	Sea basses, groupers
Callanthiidae	Rosey perches

Pseudochromidae	Dottybacks
Grammatidae	Basslets
Opistognathidae	Jawfishes
Plesiopidae	Roundheads
Notograptidae	Bearded snakeblennies
Pholidichthyidae	Convict blennies
Cepolidae	Bandfishes
Glaucosomatidae	Pearl perches
Terapontidae	Grunters
Banjosidae	Banjosids
Kuhliidae	Aholeholes
Centrarchidae	Sunfishes
Percidae	Perches, darters
Priacanthidae	Bigeyes (catalufas)
Apogonidae	Cardinalfishes
Dinolestidae	Dinolestid
Sillaginidae	Smelt whitings (whitings)
Malacanthidae	Tilefishes
Labracoglossidae	Labracoglossids
Lactariidae	False trevrelatives
Pomatomidae	Bluefish (tailor)
Menidae	Moonfish
Leiognathidae	Ponyfishes (slipmouths)
Bramidae	Pomfrets
Caristiidae	Manefishes
Arripidae	Australian salmons
Emmelichthyidae	Rovers
Lutjanidae	Snappers (emperors)
Lobotidae	Tripletails

Gerreidae	Mojarras (silver biddies)		

Gerreidae — Mojarras (silver biddies)
Haemulidae — Grunts
Inermiidae — Bonnetmouths
Sparidae — Porgies (breams)
Centracanthidae — Centracanthids
Lethrinidae — Scavengers (emperors)
Nemipteridae — Threadfin, monocle breams
Sciaenidae — Drums, croakers
Mullidae — Goatfishes
Monodactylidae — Moonfishes (fingerfishes)
Pempherididae — Sweepers
Leptobramidae — Beachsalmon
Bathyclupeidae — Bathyclupeids
Toxotidae — Archerfishes
Coracinidae — Galjoenfishes
Kyphosidae — Seachubs
Girellidae — Nibblers (blackfishes)
Scorpididae — Halfmoons, mado (sweeps)
Microcanthidae — Stripey
Ephippidae — Spadefishes
Scatophagidae — Scats
Chaetodontidae — Butterflyfishes
Pomacanthidae — Angelfishes
Enoplosidae — Oldwife
Pentacerotidae — Armorheads
Oplegnathidae — Knifejaws
Icosteidae — Ragfish
Kurtidae — Nurseryfishes
Scombrolabracidae — Scombrolabracid

Suborder Carangoidei

Nematistiidae — Roosterfish
Carangidae — Jacks (trevallies)
Echeneidae — Remoras
Rachycentridae — Cobia (black kingfish)
Coryphaenidae — Dolphins

Suborder Cirrhitoidei

Cirrhitidae — Hawkfishes
Chironemidae — Kelpfishes
Aplodactylidae — Aplodactylids
Cheilodactylidae — Morwongs
Latrididae — Trumpeters

Suborder Mugiloidei

Mugilidae — Mullets

Suborder Polynemoidei

Polynemidae — Threadfins

Suborder Labroidei

Chichlidae — Cichlids
Embiotocidae — Surfperches
Pomacentridae — Damselfishes, anemonefishes
Labridae — Wrasses
Odacidae — Rock whitings
Scaridae — Parrotfishes

Suborder Zoarcoidei

Bathymasteridae — Ronquils
Zoarcidae — Eelpouts
Stichaeidae — Pricklebacks
Cryptacanthodidae — Wrymouths
Pholididae — Gunnels
Anarhichadidae — Wolffishes
Ptilichthyidae — Quillfishes
Zaproridae — Prowfish
Scytalinidae — Graveldiver

Suborder Notothenioidei

Bovichtidae — Thornfishes
Nototheniidae — Nototheniids
Harpagiferidae — Plunderfishes
Bathydraconidae — Antarctic dragonfishes
Channichthyidae — Icefishes

Suborder Trachinoidei

Chiasmodontidae — Swallowers
Champsodontidae — Champsodontids
Trichodontidae — Sandfishes
Trachinidae — Weeverfishes
Uranoscopidae — Stargazers
Trichonotidae — Sanddivers
Creediidae — Sandburrowers
Leptoscopidae — Leptoscopids
Percophidae — Duckbills
Pinguipedidae — Sandperches
Cheimarrhichthyidae — Torrentfish
Ammodytidae — Sandlances

Suborder Blennioidei

Tripterygiidae	Triplefins
Dactyloscopidae	Sand stargazers
Labrisomidae	Labrisomids
Clinidae	Kelp blennies (kelpfishes)
Chaenopsidae	Tube blennies
Blenniidae	Combtooth blennies

Suborder Gobiodei

Rhyacichthyidae	Loach goby
Odontobutidae	Freshwater Asian gobies
Gobiidae	Gobies, sleepers (gudgeons)

Suborder Acanthuroidei

Siganidae	Rabbitfishes
Luvaridae	Louvar
Zanclidae	Moorish idol
Acanthuridae	Surgeonfishes (tangs)

Suborder Scombroidei

Sphyraenidae	Barracudas
Gempylidae	Snake mackerels, gemfishes
Trichiuridae	Cutlassfishes
Scombridae	Mackerels, tunas, bonitos
Xiphiidae	Swordfish
Istiophoridae	Billfishes

Suborder Stromateoidei

Amarsipidae	Amarsipids
Centrolophidae	Medusafishes
Nomeidae	Driftfishes
Ariommatidae	Ariommatids
Tetragonuridae	Squaretails
Stromateidae	Butterfishes

Suborder Anabantoidei

Badidae	Chameleonfish
Nandidae	Leaffishes
Pristolepidae	False leaffishes
Channidae	Snakeheads
Anabantidae	Climbing gouramies
Belontiidae	Gouramies
Helostomatidae	Kissing gouramie
Osphronemidae	Giant gouramie
Luciocephalidae	Pikehead

ORDER PLEURONECTIFORMES — FLATFISHES

Psettodidae	Spiny flatfishes
Citharidae	Citharids
Bothidae	Lefteye flounders
Pleuronectidae	Tonguefishes (tongue soles)
Soleidae	Soles
Achiridae	American soles

ORDER TETRAODONTIFORMES — TRIGGERFISHES AND RELATIVES

Triacanthodidae	Spikefishes
Triacanthidae	Triplespines
Balistidae	Triggerfishes
Monacanthidae	Filefishes (leatherjackets)
Aracanidae	Robust boxfishes
Ostraciidae	Boxfishes, cowfishes, trunkfishes
Triodontidae	Pursefish (three-toothed puffer)
Tetraodontidae	Pufferfishes (toados)
Diodontidae	Porcupinefishes
Molidae	Ocean sunfishes (molas)

Invertebrates

PHYLUM PORIFERA — SPONGES

Class Calcarea	Calcareous sponges
Class Hexactinellida	Glass sponges
Class Demospongiae	Demosponges

PHYLUM CHORDATA — INVERTEBRATE CHORDATES

Class Urochordata	Sea squirts or tunicates
Class Cephalochardata	Lancelets

PHYLUM CNIDARIA — HYDRAS, JELLYFISH, SEA ANEMONES, CORALS

Class Hydrozoa	Hydras and freshwater jellyfish
Class Scyphozoa	Jellyfish
Class Cubozoa	Jellyfish
Class Anthozoa	Sea anemones, corals, sea fans, sea pansies

PHYLUM CTENOPHORA — SEA WALNUTS OR COMB JELLIES

Class Tentaculata
Class Nuda

PHYLUM PLATYHELMINTHES — FLATWORMS

Class Trematoda	Flukes
Class Monogenea	
Class Cestoda	Tapeworms
Class Turbellaria	Free-living flatworms
Subclass Archoophora	Archoophoran turbellarians
Subclass Neoophora	Neoophoran turbellarians

PHYLUM GNATHOSTOMULIDA — ACOELOMATE WORMS

Phylum Gnathostomulida	Acoelomate worms

PHYLUM ORTHONECTIDA

Phylum Orthonectida

PHYLUM RHOMBOZOA

Class Dicyemida
Class Heterocyemida

PHYLUM NEMERTEA — RIBBON OR PROBOSCIS WORMS

Class Anopla	Unarmed nemerteans
Class Enopla	Armed nemerteans

PHYLUM GASTROTRICHA — MARINE AND FRESHWATER GASTROTRICHS

Phylum Gastrotricha	Marine and freshwater gastrotrichs

PHYLUM NEMATODA — ROUNDWORMS

Class Adenophorea
Class Secernentea

PHYLUM NEMATOMORPHA — HORSEHAIR OR HAIRWORMS

Class Nectonematoida	Marine nematomorphs
Class Gordioida	Freshwater and semiterrestrial nematomorphs

PHYLUM ROTIFERA — ROTIFERS

Class Seisonidea
Class Bdelloidea
Class Monogononta

PHYLUM ACANTHOCEPHALA — ENDOPARASITES

Class Archiacanthocephala	Parasites of birds and mammals
Class Eoacanthocephala	Parasites of fishes and reptiles
Class Palaeacanthocephala	Parasites of all vertebrates

PHYLUM KINORHYNCHA

Order (class) Cyclorhagida
Order (class) Homalorhagida

PHYLUM LORICIFERA

Phylum Loricifera

PHYLUM TARDIGRADA   WATER BEARS

Class Heterotardigrada
Class Eutardigrada

PHYLUM SIPUNCULA   PEANUT WORMS

Phylum Sipuncula   Peanut worms

PHYLUM ECHIURA   SPOONWORMS

Phylum Echiura   Spoonworms

PHYLUM PRIAPULIDA

Phylum Priapulida

PHYLUM MOLLUSCA   MOLLUSKS

Class Aplacophora	Wormlike mollusks
Class Polyplacophora	Chitons
Class Monoplacophora	
Class Gastropoda	Gastropods
Class Bivalvia	Bivalves
Class Scaphopoda	
Class Cephalopoda	Squid and octopi

PHYLUM ANNELIDA   SEGMENTED WORMS

Class Polychaeta
Class Oligochaeta
Class Hrudinea   Leeches
Class Branchiobdellida

PHYLUM POGONOPHORANS

Phylum Pogonophorans

PHYLUM ARTHROPODA   ARTHROPODS

Class Insecta	Insects
Subclass Entognatha	Wingless insects
Subclass Ectognatha	Winged insects

SUBPHYLUM CHELICERATA

Class Merostomata
Class Arachnida   Spiders, scorpions, mites and ticks
Class Pycnogonida   Sea spiders

SUBPHYLUM CRUSTACEA   CRUSTACEANS

Class Remipedia	
Class Cephalocarida	
Class Branchiopoda	
Class Ostracoda	Ostracods
Class Copepoda	
Class Mystacocarida	
Class Tantulocarida	
Class Branchiura	
Class Cirripedia	Barnacles
Class Pentastomida	
Class Malacostraca	Crabs, lobsters, shrimp

SUBPHYLUM MYRIAPODA   MYRIAPODUS ARTHROPODS

Class Chilopoda	Centipedes
Class Symphyla	
Class Diplopoda	Millipedes
Class Pauropoda	Pauropods

PHYLUM ONYCHOPHORA

Phylum Onychophora

PHYLUM HEMICHORDATA

Class Enteropneusta   Acorn worms
Class Pterobranchia

PHYLUM CHORDATA

SUBPHYLUM UROCHORDATA

Class Ascidiacea   Sea squirts
Class Thaliacea
Class Larvacea

SUBPHYLUM
CEPHALOCHORDATA
LANCELETS

PHYLUM CHAETOGNATHA ARROWWORMS

Phylum Chaetognatha   Arrowworms

PHYLUM
ECHINODERMATA
ECHINODERMS

SUBPHYLUM HOMALOZOA

SUBPHYLUM CRINOZOA

Class Crinoidea   Sea lilies, feather stars

SUBPHYLUM ASTEROZOA

Class Asteroidea   Sea stars
Class Ophiuroidea   Brittle stars
Class Concentricycloidea

SUBPHYLUM ECHINOZOA

Class Echinoidea   Sea urchins, sand dollars
Class Holothuroidea   Sea cucumbers

PHYLUM BRYOZOA

Class Phylactolaemata
Class Stenolaemata
Class Gymnolaemata

PHYLUM ENTOPROCTA

Phylum Entoprocta

PHYLUM PHORONIDA

Phylum Phoronida

PHYLUM BRACHIOPODA   BRACHIPODS

Class Inarticulata
Class Articulata

CLASS ARACHNIDA

ORDER ARANEAE   SPIDERS

Agelenidae	Funnel weavers
Antrodiaetidae	Folding trapdoor spiders
Araneidae	Orb weavers
Clubionidae	Sac spiders
Ctenidae	Wandering spiders
Ctenizidae	Trapdoor spiders
Dictynidae	Dictynid spiders
Dipluridae	Funnel-web spiders
Eresidae	Eresid spiders
Heteropodidae	Huntsman spiders
Linyphiidae	Dwarf spiders
Loxoscelidae	Violin spiders
Lycosidae	Wolf spiders
Oxyopidae	Lynx spiders
Pholcidae	Daddy-long-legs spiders
Pisauridae	Nursery-web spiders
Salticidae	Jumping spiders
Scytodidae	Spitting spiders
Selenopidae	Selenopid crab spiders
Sparassidae	Giant crab spiders
Tetragnathidae	Large-jawed orb weavers
Theraphosidae	Tarantulas
Theridiidae	Comb-footed spiders
Thomisidae	Crab spiders

ORDER SCORPIONES   SCORPIONS

Buthidae	Buthid scorpions
Iuridae	Iurid scorpions

ORDER
PSEUDOSCORPIONES
PSEUDO-SCORPIONS

Cheliferidae	
Chernetidae	Chernetids

ORDER OPILIONES   HARVESTMEN

Leiobunidae
Phalangiidae

575

ORDER ACARINA — MITES AND TICKS

Argasidae	Soft ticks
Hydrachnellae	Water mites
Ixodidae	Hard ticks
Tetranychidae	Spider mites
Trombidiidae	Velvet mites

ORDER UROPYGI — WHIPSCORPIONS

Thelyphonidae	Vinegaroons

ORDER AMBLYPYGI — TAILLESS WHIPSCORPIONS

Phrynidae	
Tarantulidae	

ORDER SOLIFUGAE — WINDSCORPIONS

Eremobatidae	
Solpugidae	

CLASS INSECTA

ORDER PROTURA — PROTURANS

Eosentomidae	

ORDER COLLEMBOLA — SPRINGTAILS

Entomobryidae	
Hypogastruridae	
Isotomidae	
Onychiuridae	
Sminthuridae	Globular springtails

ORDER DIPLURA — DIPLURANS

Campodeidae	Campodeids
Japygidae	Japygids

ORDER ARCHAEOGNATHA — BRISTLETAILS

Machilidae	Jumping bristletails
Meinertellidae	

ORDER THYSANURA — SILVERFISH

Lepismatidae	Silverfish (firebrats)

ORDER EPHEMEROPTERA — MAYFLIES

Baetidae	Small mayflies
Ephemerellidae	Midboreal mayflies
Ephemeridae	Burrowing mayflies
Heptageniidae	Stream mayflies
Leptophlebiidae	Spinners

ORDER ODONATA — DRAGONFLIES AND DAMSELFLIES

Aeschnidae	Darners
Calopterygidae	Broad-winged damselflies
Coenagrionidae	Narrow-winged damselflies
Cordulegastridae	Biddies
Corduliidae	Green-eyed skimmers
Gomphidae	Clubtails
Lestidae	Spread-winged damselflies
Libellulidae	Common skimmers
Macromiidae	Belted and river skimmers
Petaluridae	Graybacks

ORDER BLATTODEA — COCKROACHES

Blaberidae	
Blattellidae	
Blattidae	Common cockroaches

ORDER MANTODEA — MANTIDS

Empusidae	
Hymenopodidae	Flower mantids
Mantidae	Common praying mantids

ORDER ISOPTERA — TERMITES

Hodotermitidae	Rotten-wood termites
Kalotermitidae	Damp-wood termites
Mastotermitidae	
Rhinotermitidae	Subterranean termites
Termitidae	Nasutiform termites

ORDER ZORAPTERA

ZORAPTERANS

Zorotypidae — Zorapterans

ORDER GRYLLOBLATTODEA ICE INSECTS

Grylloblattidae — Rock crawlers

ORDER DERMAPTERA EARWIGS

Chelisochidae — Black earwigs
Forficulidae — Common earwigs
Labiduridae — Long-horned earwigs
Labiidae — Little earwigs

ORDER PLECOPTERA STONEFLIES

Capniidae — Small winter stoneflies
Chloroperlidae — Green stoneflies
Isoperlidae — Green-winged stoneflies
Leuctridae — Rolled-winged stoneflies
Nemouridae — Spring stoneflies
Peltoperlidae — Roachlike stoneflies
Perlidae — Common stoneflies
Perlodidae — Perlodid stoneflies
Pteronarcidae — Giant stoneflies
Taeniopterygidae — Winter stoneflies

ORDER ORTHOPTERA GRASSHOPPERS AND CRICKETS

Acrididae — Short-horned grasshoppers
Cooloolidae — Cooloola monster
Cylindrachetidae — Sand gropers
Eneopterinae — Bush crickets
Eumastacidae; Tanaoceridae — Monkey grasshoppers
Gryllacrididae; Rhaphidophoridae — Leaf-rolling crickets, camel crickets, cave crickets
Gryllidae — True crickets
Gryllotalpidae — Mole crickets
Myrmecophilidae — Ant crickets
Oecanthinae Pyrgomorphidae — Tree crickets
Stenopelmatidae — King crickets
Tetrigidae — Pygmy grasshoppers
Tettigoniidae — Long-horned grasshoppers and katydids
Tridactylidae — Pygmy mole crickets

ORDER PHASMATODEA STICK AND LEAF INSECTS

Phasmatidae — Stick insects
Phasmidae — Walking sticks
Timemidae — Timemas

ORDER EMBIOPTERA WEBSPINNERS

Clothodidae — Clothodids
Embiidae — Embiids

ORDER PSOCOPTERA BOOKLICE AND BARKLICE

Lepidopsocidae — Scaly barklice
Liposcelididae — Liposcelid booklice
Pseudocaeciliidae
Psocidae — Common barklice
Psyllipsocidae — Psyllipsocids
Trogiidae — Trogiid booklice

ORDER PHTHIRAPTERA PARASITIC LICE

Boopidae
Echinophthiriidae — Spiny sucking lice
Gyropidae — Guinea pig lice
Haematopinidae — Mammal-sucking lice
Hoplopleuridae
Laemobothriidae — Bird lice
Linognathidae — Smooth sucking lice
Menoponidae — Poultry-chewing lice
Pediculidae — Human lice
Philopteridae — Feather-chewing lice
Ricinidae — Bird lice
Trichodectidae — Mammal-chewing lice

ORDER HEMIPTERA BUGS

Achilidae — Achilid planthoppers
Adelgidae — Pine aphids
Aleyrodidae — Whiteflies
Alydidae — Broad-headed bugs
Anthocoridae — Flower bugs, minute pirate bugs
Aphididae Aphrophoridae — Aphids
Aradidae — Flat bugs, bark bugs
Asterolecaniidae — Pit scales
Belostomatidae — Giant water bugs

Berytidae	Stilt bugs
Carsidaridae	
Cercopidae	Spittlebugs and froghoppers
Chermidae	Pine and spruce aphids
Cicadellidae	Leafhoppers
Cicadidae	Cicadas
Cimicidae	Bed bugs
Cixiidae	Cixiid planthoppers
Coccidae; Colobathristidae	Soft scales, wax scales
Coreidae	Leaf-footed bugs, crusader bugs
Corixidae	Water boatmen
Cydnidae	Negro bugs
Dactylopiidae	Cochineal bugs
Delphacidae	Delphacid planthoppers
Derbidae	Derbid planthoppers
Diaspididae	Armored scale insects
Dictyophdridae; Dinidoridae	Dictyopharid planthoppers
Dipsocoridae; Schizopteridae	Jumping ground bugs
Eriosomatidae	Woolly and gall-making aphids
Eurymelidae; Membracidae	Treehoppers
Flatidae	Flatid planthoppers
Fulgoridae	Fulgorids
Gelastocoridae	Toad bugs
Gerridae	Water striders
Hebridae Homotomidae	Velvet water bugs
Hydrometridae	Water measurers
Isometopidae	Jumping tree bugs
Issidae	Issid planthoppers
Kermidae	Gall-like coccids
Kerriidae; Lacciferidae	Lac insects
Leptopodidae	Spiny shore bugs
Lygaeidae	Seed bugs
Margarodidae	Giant scale insects
Mesoveliidae	Water treaders
Miridae	Leaf or plant bugs
Nabidae	Damsel bugs
Naucoridae	Creeping water bugs
Nepidae	Waterscorpions
Notonectidae	Backswimmers
Ochteridae	Velvety shore bugs
Ortheziidae	Ensign scales
Peloridiidae	Moss bugs
Pentatomidae	Stink bugs, shield bugs
Phylloxeridae	Gall aphids
Phymatidae	Ambush bugs

Polyctenidae	Bat bugs
Pseudococcidae; Eriococcidae	Mealybugs
Psyllidae	Psyllids, lerps
Pyrrhocoridae	Red bugs, stainers
Reduviidae	Assassin bugs
Rhopalidae	Scentless plant bugs
Ricaniidae	Ricaniid planthoppers
Saldidae	Shore bugs
Scutelleridae	Jewel bugs, shield-backed bugs
Tessaratomidae	
Tettigarctidae	Hairy cicadas
Tingidae	Lace bugs
Triozidae	
Veliidae	Ripple bugs

ORDER THYSANOPTERA THRIPS

Aeolothripidae	Banded thrips
Phloeothripidae	Tube-tailed thrips
Thripidae	Common thrips

ORDER MEGALOPTERA ALDERFLIES AND DOBSONFLIES

Corydalidae	Dobsonflies
Sialidae	Alderflies

ORDER RAPHIDIOPTERA SNAKEFLIES

Inocelliidae	
Raphidiidae	

ORDER NEUROPTERA NET-VEINED INSECTS

Ascalaphidae	Owlflies
Chrysopidae	Green lacewings
Coniopterygidae	Dusty-wings
Hemerobiidae	Brown lacewings
Ithonidae	Moth lacewings
Mantispidae	Mantidflies
Myrmeleontidae	Antlions
Nemopteridae	
Polystoechotidae	Giant lacewings
Psychopsidae	Silky lacewings
Sisyridae	Spongeflies

ORDER COLEOPTERA — BEETLES

Anobiidae	Furniture beetles
Anthribidae	Fungus weevils
Bostrichidae	Branch and twig borers, auger beetles
Brentidae	Primitive weevils
Bruchidae	Seed beetles
Buprestidae	Metallic wood-boring beetles, jewel beetles
Cantharidae	Soldier beetles
Carabidae	Ground beetles
Cerambycidae	Longhorn beetles, longicorn beetles
Chrysomelidae	Leaf beetles
Cicindelidae	Tiger beetles
Cleridae	Checkered beetles
Coccinellidae	Ladybird beetles
Cucujidae	Flat bark beetles
Curculionidae	Snout beetles and weevils
Dermestidae	Dermestid beetles
Dytiscidae	Predacious diving beetles
Elateridae	Click beetles
Erotylidae	Pleasing fungus beetles
Gyrinidae	Whirligig beetles
Haliplidae	Crawling water beetles
Histeridae	Hister beetles
Hydrophilidae; Laemophloeidae	Water scavenger beetles
Lampyridae	Fireflies, lightning bugs
Lathridiidae	Minute brown scavenger beetles
Lucanidae	Stag beetles
Lycidae	Net-winged beetles
Lymexylidae	Ship-timber beetles
Meloidae	Blister beetles
Melyridae	Softwinged flower beetles
Mycetophagidae	Hairy fungus beetles
Nitidulidae	Nitidulid beetles
Passalidae	Passalid beetles or bessbugs
Psephenidae	Water pennies
Ptiliidae	Feather-winged beetles
Ptinidae	Spider beetles
Pyrochroidae	Fire-colored beetles
Rhipiphoridae	Rhipiphoridan beetles
Scarabaeidae	Scarab beetles
Scolytidae	Bark and ambrosia beetles
Silphidae	Carrion beetles
Silvanidae	Flat grain beetles
Staphylinidae	Rove beetles
Tenebrionidae	Darkling beetles
Trogidae	Carcass beetles
Trogositidae	Bark-gnawing beetles

ORDER STREPSIPTERA — TWISTED-WINGED PARASITES

Mengeidae	
Stylopidae	Stylopids

ORDER MECOPTERA AND RELATIVES — SCORPIONFLIES

Bittacidae	Hangingflies
Boreidae	Snow scorpionflies
Panorpidae	Common scorpionflies

ORDER SIPHONAPTERA — FLEAS

Dolichopsyllidae; Ceratophyllidae	Rodent fleas
Leptopsyllidae	Mouse fleas
Pulicidae	Common fleas
Tungidae	Sticktight and chigoe fleas

ORDER DIPTERA — FLIES

Acroceridae	Bladder flies
Agromyzidae	Leafminer flies
Anthomyiidae	Anthomyiid flies
Apioceridae	Flower-loving flies
Asilidae	Robber flies
Bibionidae	March flies
Blephariceridae	Net-winged midges
Bombyliidae	Bee flies
Braulidae	Beelice
Calliphoridae	Blowflies
Canacidae	Beach flies
Cecidomyiidae	Gall midges
Ceratopogonidae	Punkies, biting midges
Chaoboridae	Phantom midges
Chironomidae	Midges
Chloropidae	Grass flies

Coelopidae	Seaweed flies
Conopidae	Thick-headed flies
Culicidae	Mosquitoes
Dolichopodidae	Long-legged flies
Drosophilidae	Vinegar (pomace) flies
Empididae	Dance flies, water cruisers
Ephydridae	Shore flies
Fanniidae	
Fergusoninidae	Eucalyptus flies
Heleomyzidae	Sun flies
Hippoboscidae	Louse flies and sheep keds
Lonchaeidae	Lance flies
Micropezidae	Stilt-legged flies
Muscidae	House flies and bushflies
Mycetophilidae	Fungus gnats
Mydidae	Mydas flies
Nemestrinidae	Tanglevein flies
Nerriidae	Cactus flies
Neurochaetidae	Upside-down flies
Nycteribiidae; Streblidae	Bat flies
Oestridae; Gasterophilidae	Botflies
Phoridae	Scuttle flies
Piophilidae	Skipper flies
Platypezidae	Flat-footed flies
Platystomatidae	Platystomatid flies
Psychodidae	Moth flies
Ptychopteridae	Phantom crane flies
Pyrgotidae	Pyrgotid flies
Rhagionidae	Snipe flies
Sarcophagidae	Flesh flies
Scenopinidae	Window flies
Sciaridae	Black fungus gnats
Sciomyzidae	March flies
Sepsidae	Ant flies
Simuliidae	Black or sand flies
Sphaeroceridae	Short heel flies, small dung flies
Stratiomyidae	Soldier flies
Syrphidae	Hover flies
Tabanidae	Horse and deer flies
Tachinidae	Tachinid flies
Tephritidae	Fruit flies
Teratomyzidae	Fern flies
Therevidae	Stiletto flies
Tipulidae	Crane flies
Trichoceridae	Winter crane flies

ORDER TRICHOPTERA   CADDISFLIES

Hydropsychidae	Net-spinning caddisflies
Leptoceridae	Long-horned caddisflies
Limnephilidae	Northern caddisflies
Phryganeidae	Large caddisflies
Psychomyiidae	Tube-making caddisflies

ORDER LEPIDOPTERA   BUTTERFLIES AND MOTHS

Alucitidae	Many-plume moths
Anthelidae	
Apaturidae	Hackberry and goatweed butterflies
Arctiidae	Tiger moths
Bombycidae	Silkworm moths
Bucculatricidae	
Carposinidae	Carposinid moths
Carthaeidae	
Castniidae	
Citheroniidae	Royal moths
Coleophoridae	Casebearer moths
Cosmopterygidae	
Copromorphidae	
Cossidae	Wood moths
Ctenuchidae	Ctenuchid moths
Danaidae	Milkweed butterflies
Drepanidae	Hook-tip moths
Eupterotidae; Zanolidae	Zanolid moths
Gelechiidae	Gelechiid moths
Geometridae	Measuringworm moths, loopers
Gracilariidae	Leaf blotch miners
Hepialidae	Ghost moths, swifts
Hesperiidae	Skippers
Incurvariidae	Yucca moths and relatives
Lasiocampidae	Tent caterpillar and lappet moths
Libytheidae	Snout butterflies
Limacodidae	
Lycaenidae	Gossamer-winged butterflies
Lymantriidae	Tussock moths and relatives
Lyonetiidae	Lyonetiid moths
Noctuidae	Noctuids
Notodontidae	Prominents
Nymphalidae	Brush-foots

Oecophoridae	Oecophorid moths
Papilionidae	Swallowtails and relatives
Pieridae	Whites and sulfurs
Plutellidae	Diamondback moths
Pterophoridae	Plume moths
Psychidae	Bagworm moths
Pyralidae	Pyralid moths
Riodinidae	Metalmarks
Saturniidae	Giant silkworm moths

Satyridae	Satyrs, nymphs, and arctics
Sesiidae	Clear-winged moths
Sphingidae	Sphinx or hawk moths
Thaumetopoeidae	
Tineidae	Clothes moths and relatives
Tortricidae	Tortricid moths
Uraniidae	
Yponomeutidae	Ermine moths
Zygaenidae	Smoky moths

ORDER HYMENOPTERA BEES, ANTS, WASPS AND SAWFLIES

Agaonidae	Fig wasps
Andrenidae	Andrenid bees
Apidae	Digger bees, carpenter bees, cuckoo bees, bumblebees, honeybees and relatives
Argidae	Argid sawflies
Braconidae	Braconids
Cephidae	Stem sawflies
Chalcididae	Chalcids
Chrysididae	Cuckoo wasps
Cimbicidae	Cimbicid sawflies
Colletidae	Yellow-faced and plasterer bees
Cynipidae	Gall wasps
Encyrtidae	Encyrtid wasps
Eulophidae; Aphelininae	Eulophid wasps
Eurytomidae	Seed chalcids
Evaniidae	Hatchet wasps
Formicidae	Ants
Halictidae	Sweat (halictid) bees
Ibaliidae	Ibaliid wasps

Ichneumonidae	Ichneumonids
Megachilidae	Leaf-cutting bees
Melittidae	Melittid bees
Mutillidae	Velvet ants
Mymaridae	Fairy wasps
Pelecinidae	Pelecinids
Pergidae	Pergid sawflies
Pompilidae	Spider wasps
Pteromalidae	Pteromalid wasps
Scelionidae	Scelionid wasps
Scoliidae	Scoliid wasps
Siricidae	Horntails
Sphecidae	Sphecid wasps
Stephanidae	Stephanid wasps
Symphyta	Sawflies
Tenthredinidae	Common sawflies
Tiphiidae	Tiphid wasps
Torymidae	Torymid wasps
Vespidae	Vespid wasps

Size Comparisons

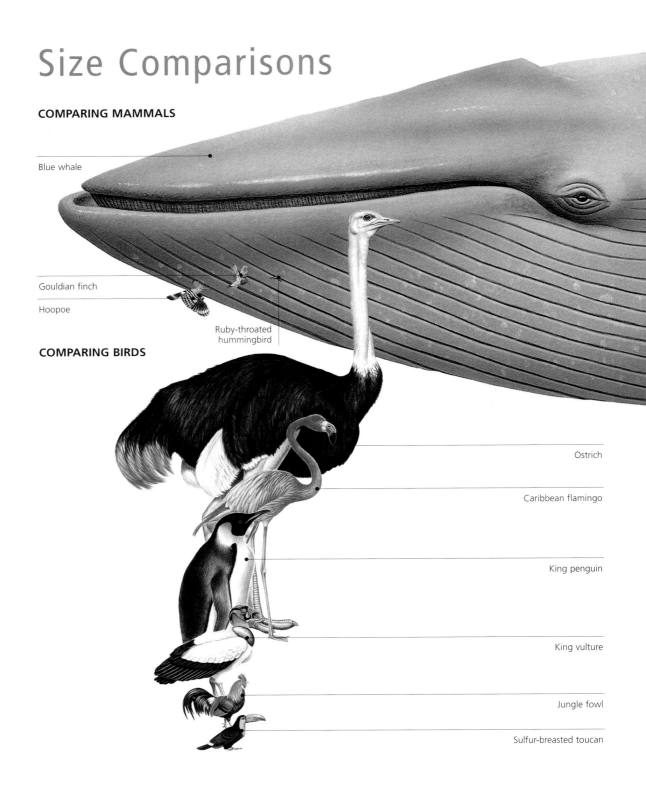

COMPARING MAMMALS

Blue whale

Gouldian finch

Hoopoe

Ruby-throated
hummingbird

COMPARING BIRDS

Ostrich

Caribbean flamingo

King penguin

King vulture

Jungle fowl

Sulfur-breasted toucan

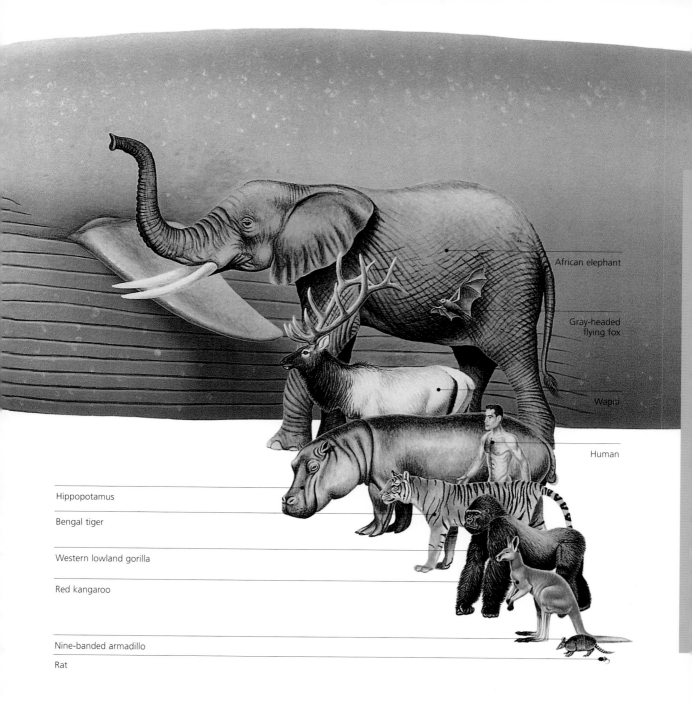

African elephant

Gray-headed
flying fox

Wapiti

Human

Hippopotamus

Bengal tiger

Western lowland gorilla

Red kangaroo

Nine-banded armadillo

Rat

Size Comparisons

COMPARING REPTILES

Indopacific crocodile

Green anaconda

COMPARING FISHES

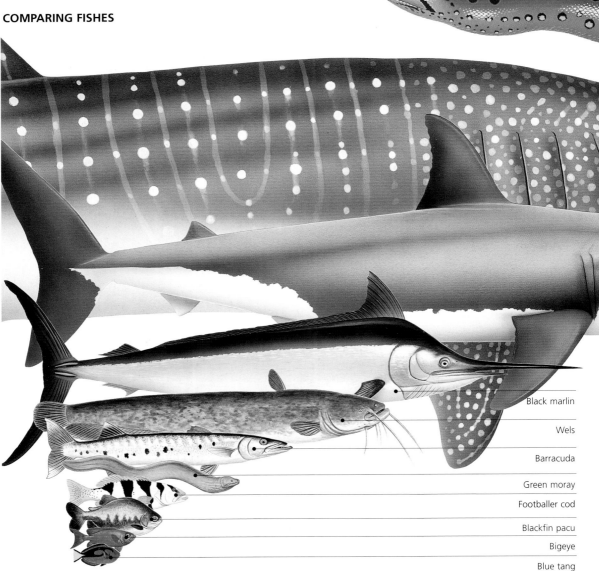

Black marlin

Wels

Barracuda

Green moray

Footballer cod

Blackfin pacu

Bigeye

Blue tang

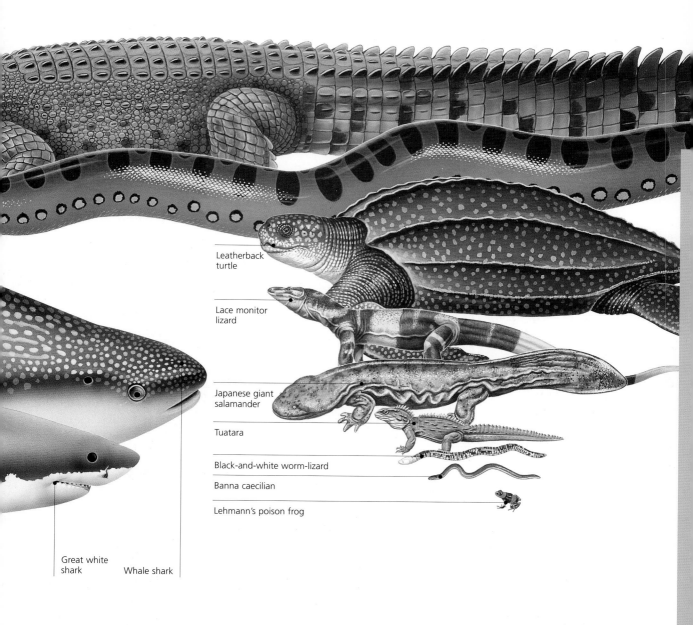

Leatherback
turtle

Lace monitor
lizard

Japanese giant
salamander

Tuatara

Black-and-white worm-lizard

Banna caecilian

Lehmann's poison frog

Great white
shark   Whale shark

Index

C

Acknowledgments

Weldon Owen would like to thank the following people for their assistance in the production of this book: Philippa Findlay and Sarah Plant at Puddingburn Editorial Services, Andrew Davies at Creative Communications, Brendan Cotter, Helen Flint, Grace Newell and Guy Troughton.

The following illustrations are © MagicGroup s.r.o. (Czech Republic) www.magicgroup.cz:
(t=top, tl=top left, tc=top center, tr=top right, cl=center left, c=center, cr=center right, b=bottom, bl=bottom left, bc=bottom center, br=bottom right).

78tc, 79tr, 92bl, 94tr, 102tr, 106t,cl, 108tl,bl,br, 109t,tr,br, 111cl,cr, 112c,bl, 113tl,cr,bl,br, 114t,cl, 115tr,bl, 116t,cl, 117br, 138t,bl,br, 139tr,cl,cr, 140c,b, 141tl,tr,cl,cr,bl,bc,br, 159tc, 160tr,c,bl,br, 161tl,tr, 178tc, 180b, 181bl,br, 186b, 190cr, 201c, 205tc, 206tr,c,bl, 207tl,tr,c,bc, 212c,b, 213t,b, 215tl,tr,cl,br,bl, 216tl,tc, 217c,br, 223tl, 224cl,c,cr, 225c,bl,br, 226br, 227tl, 230tr,cl,c,cr,bl,br, 231tl,tr, 237tr, 249cl,br, 250cl, 270br, 274tr, 276c, 278tr,cl,cr, 281bl, 291br, 292cl,bl, 293br, 294bl,bc,br, 295br, 298tr,cl,c,br, 299tr,cl, 301tc, 302tr,br, 303cl, 304c, 305tl,tr,c, 307tl,cr, 308tr,cr, 311cl, 313tl,c,bl,br, 314t, 320cl, 321bl,br, 332br, 334cl,bl,br, 366bl, 368tr,cl,cr, 370b, 371b, 373b, 374c, 375t, 377br, 378cl, 382br, 383t,tr,bl, 392cl, 400tl,tc, 401tc,tr, 402tr, 404cl, 405c, 406tr, 407b, 408cl, 409cl,bl,br, 410tl,tr, 412tr, 413cl,c, 414t,cl,br, 415br, 416tr, 417tr,br, 426tc,br, 427tl,tc,tr,bl,bc, 432tr,c,bc, 448tl,tc,tr, 452tr,c, 453tc,c, 454tr,bc, 455bc, 458tr,cl,cr,bl,br, 459tc,c, 460tr,cl,bc,br, 461bc, 464tr,bc, 466tl,tr,c,bc, 467bc, 469tl, 470tr, 471bl,br, 476cl, 491tl, 493tl,tc,tr, 496cl, 500bl,br, 501tl, 514tl,tr,bl, 565cr, 566tc, 567cr, 568tr,c

Illustrators: Susanna Addario, Art Studio, Mike Atkinson/Garden Studio, Paul Bachem, Alistair Barnard, Priscilla Barret/Wildlife Art Ltd., Jane Beatson, Sally Beech, Andre Boos, Anne Bowman, Peter Bull, Martin Camm, Stuart Carter/Wildlife Art Ltd., D. Cole/Wildlife Art Ltd, Marjorie Crosby-Fairall, B. Croucher/Wildlife Art Ltd., Marc Dando/Wildlife Art Ltd., Peter David, Fiammetta Dogi, Sandra Doyle/Wildlife Art Ltd., Gerald Driessen, Simone End, Christer Erikson, Alan Ewart, Lloyd Foye, John Francis/Bernard Thornton Artists UK, Peg Gerrity, John Gittoes, Mike Golding, Mike Gorman, Ray Grinaway, Gino Hasler, Tim Hayward/ Bernard Thornton Artists UK, Dr. Stephen Hutchinson, Robert Hynes, Ian Jackson/Wildlife Art Ltd., Janet Jones, Roger Kent, David Kirshner, Frank Knight, Angela Lober, Frits Jan Maas, John Mac/FOLIO, David Mackay, Martin Macrae/FOLIO, Rob Mancini, Karel Mauer, Iain McKellar, James McKinnon, David Moore/Linden Artists, Robert Morton, Colin Newman/Bernard Communications, Ken Oliver/Wildlife Art Ltd., Erik van Ommen, Nicola Oram, Photodisc, Maurice Pledger, Tony Pyrzakowski, Oliver Rennert, John Richards, Edwina Riddell, Steve Roberts/Wildlife Art Ltd., Barbara Rodanska, Trevor Ruth, Claudia Saraceni, R.T. Sauey, Michael Saunders, Peter Schouten, Peter Scott/Wildlife Art Ltd., Rod Scott, Chris Shields/Wildlife Art Ltd., Ray Sim, Marco Sparaciari, Chris Stead, Kevin Stead, Mark Stewart, Stockbyte, Roger Swainston, Steve Trevaskis, Thomas Trojer, Guy Troughton, Chris Turnbull/Wildlife Art Ltd., Glen Vause, Genevieve Wallace, Trevor Weekes, Rod Westblade, Ann Winterbotham.